QUANTUM MECHANICS
FOR ORGANIC CHEMISTS

HOWARD E. ZIMMERMAN

Chemistry Department
University of Wisconsin
Madison, Wisconsin

ACADEMIC PRESS   New York   San Francisco   London   1975

A Subsidiary of Harcourt Brace Jovanovich, Publishers

ACADEMIC PRESS, INC.
111 Fifth Avenue, New York, New York 10003

United Kingdom Edition published by
ACADEMIC PRESS, INC. (LONDON) LTD.
24/28 Oval Road, London NW1

Library of Congress Cataloging in Publication Data

Zimmerman, Howard E
  Quantum mechanics for organic chemists.

  Includes bibliographical references and index.
  1. Quantum chemistry.   2.   Chemistry, Physical
organic.   I.   Title.
QD462.Z55   530.1′2′024541   74-17971
ISBN 0–12–781651–8

PRINTED IN THE UNITED STATES OF AMERICA

To Jane

CONTENTS

PREFACE

During the last century organic chemistry has developed far from its empirical beginnings, where each observation constituted an isolated fact. As a result of a century's accumulation of experimental fact, generalizations have evolved. Most have proven understandable from a qualitative theoretical viewpoint. This framework of generalization interwoven with qualitative theory constitutes the basis of the so-called organic chemist's intuition. Organic intuition is a powerful tool not yet formulated on a quantitative basis.

Part of the theoretical framework has been resonance, or qualitative valence bond, reasoning. Thus the structures of organic molecules are represented in valence bond symbolism, and such representation is efficaciously used in formulating reactions and reactivity. Following three decades of the use of this approach in organic chemistry, it now appears that there is much to be gained from the use of molecular orbital theory in formulating organic systems. In certain instances where the resonance approach is unsatisfactory, molecular orbital theory can provide acceptable rationalization of facts and the prediction of new ones. Even the most simplistic of molecular orbital methods lead to molecular conclusions and provide insight which are in remarkable agreement with experimental observation. More generally, a look at the same phenomenon from this second viewpoint is refreshing.

As for the level of sophistication required in this approach, one must consider the following aspects. For practicality, there is no point in complexity exceeding need. In fact, resonance reasoning has had such a profound impact on organic chemistry because it can be used quickly and conveniently. Often where one desires only a prediction of the order of reactivity of a set of reactants or of different sites in one reactant or where one wishes insight into the source of some peculiar pattern of experimental results, the simpler methods of quantum mechanics such as the Hückel theory suffice. When this level of sophistication is inadequate, then it is necessary to employ more sophisticated methods. Even where the simplest (e.g. Hückel) theory predicts molecular behavior correctly, it is worthwhile for the interested organic chemist to penetrate further into more sophisticated quantum mechanics. A most intriguing endeavor is the comparison of prediction by various levels of sophistication with one another and with experimental reality.

In the past, quantum mechanics was a difficult field for organic chemists Some excellent texts have, however, become available during the last decade. The present text is based on the author's lecture notes used at Wisconsin since 1960 in a first-year graduate course for organic majors and also used in a series of American Chemical Society "Short Courses." The treatment begins on a very elementary level and proceeds through the Hückel level into more advanced methods such as polyelectron theory. The transition from elementary to advanced material is purposely gradual. Often a given topic is reconsidered several times with equivalent but alternative treatments. Where possible, the organic chemist is taught the language and its use prior to its theoretical justification. It is the author's experience that this approach develops in the student an intuitive feeling for the "how" and "why" of quantum mechanics. A number of items not appearing elsewhere are included. The pedagogical method employed is that of gradual escalation of difficulty while maintaining the organic chemist's interest. It is the author's hope that the reader will experience the joy in learning an intriguing subject unimpeded by an often justified concern in succeeding.

Acknowledgment

Some word is needed to acknowledge the stimulation afforded by my research students. This dedicated group of individuals has provided a spirited and stimulating atmosphere with unusual intellectual depth and has made this effort most enjoyable.

Chapter 1

DELINEATION OF THE METHODS AND RESULTS OF THE LCAO-MO HÜCKEL APPROACH

1.1 Some Preliminary Basic Definitions and Introductory Material

1.1a Orbitals

An *orbital* is the locus in space in which an electron is distributed. Qualitatively, it may be pictured as an electron cloud. An orbital localized at one atom is called an *atomic orbital* while one which is distributed around a molecular framework is termed a *molecular orbital*. Each orbital, atomic or molecular, has two characteristics of particular interest, its energy and its spatial description.

By the *energy* of an orbital, atomic or molecular, is actually implied the energy of an electron spread about this orbital. An orbital will be seen to have mathematical significance but no physical reality until it is occupied by an electron.

The spatial description of an orbital can be qualitatively and pictorially indicated by the drawing of a surface encompassing some large fraction of the electron population of the orbital; this has been common in organic chemistry. Thus carbon $2s$ and $2p$ orbitals are depicted in Fig. 1.1-A. However, an orbital is more precisely formulated as a mathematical function of position in some coordinate system. If x, y, z coordinates are used, the following formulations of $2s$, $2p_x$, $2p_y$, and $2p_z$ orbitals (χ_{2s}, χ_{2px}, χ_{2py}, χ_{2pz}, respectively), termed Slater orbitals,[1-5] may be given:

$$\chi_{2s} = \frac{k^{5/2}}{(3\pi)^{1/2}} \rho e^{-k\rho} \qquad \text{1.1-1a}$$

1

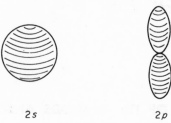

$2s$ $2p$

F $_{\mathrm{IG}}$. 1.1-A

$$\chi_{2py} = \frac{k^{5/2}}{\pi^{1/2}}\, ye^{-k\rho} \qquad\qquad\qquad 1.1\text{-}1b$$

$$\chi_{2px} = \frac{k^{5/2}}{\pi^{1/2}}\, xe^{-k\rho} \qquad\qquad\qquad 1.1\text{-}1c$$

$$\chi_{2pz} = \frac{k^{5/2}}{\pi^{1/2}}\, ze^{-k\rho} \qquad\qquad\qquad 1.1\text{-}1d$$

where $\rho = (x^2 + y^2 + z^2)^{1/2}$ is the distance of a given point in space under consideration from the origin; the origin is taken as the center of the (e.g.) carbon atom.* In the above $k = 1.625$ when ρ is in angstroms and the atom is carbon.

What we are saying is that an orbital is nothing other than a function of space coordinates x, y, and z; the value of the orbital at each point P in space will depend on the magnitude of x, y, and z at P. It is instructive to analyze the spatial characteristics of the four Slater atomic orbitals given in Eqs. 1.1-1.

The case of atomic orbital $2s$ is simplest. The Slater orbital function χ_{2s} (note Eq. 1.1-1a) has no directionality; that is, χ_{2s} is weighted equally in x, y, and z. χ_{2s}, being a simple exponential, is positive everywhere in space (Fig. 1.1-B). This is the first spatial characteristic to be noted. Second, χ_{2s} will have the same value at all points of equal distance ρ from the atomic nucleus (i.e., the origin), and hence χ_{2s} has spherical symmetry. Finally, we note that χ_{2s} is a constant times a negative exponential and we know that a negative exponential function decreases rapidly to zero as the variable taken negatively in the exponent increases. Thus as ρ, the distance from the nucleus, increases, the value of the orbital χ_{2s} decreases. From these three observations we can depict χ_{2s} pictorially as in Fig. 1.1-B. Here the everywhere positive value of χ_{2s} is noted, the spherical

---

* The reasons for the exact formulation of the constant preceding the exponent are given later in another connection.

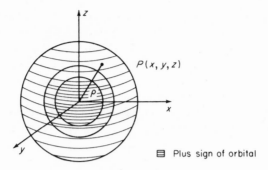

□ Plus sign of orbital

FIG. 1.1-B

symmetry is indicated, and the increasing diffuseness of the orbital at increasing distances outward is suggested.

One may now pick one of the three $2p$ atomic orbitals, as given in Eqs. 1.1-1, and inspect its characteristics. In looking at χ_{2pz}, we note that it is a product of the coordinate z (of the point P under consideration) and the negative exponential term $e^{-k\rho}$ discussed above in connection with the $2s$ orbital. The following observations may be made in sequence:

(a) Since the exponential term is positive at all points in space, the sign of χ_{2pz} will be that of z, the distance of the point P under consideration from the horizontal xy plane. Therefore, χ_{2pz} will be positive above the (horizontal) xy plane, zero in the xy plane, and negative below this plane (cf. Fig. 1.1-C).

(b) Next we can see that as the distance from the origin increases to infinity in the plus or minus z direction the orbital χ_{2pz} approaches zero. The z component (i.e., its absolute value) increases while the $e^{-k\rho}$ decreases

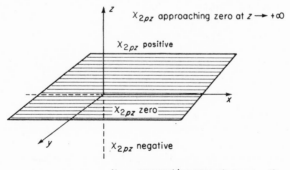

FIG. 1.1-C. Sign characteristics of p_z orbital.

in proceeding upward or downward from the center of the atom, but the negative exponential is a "stronger function" and dominates.*

(c)   Third, since χ_{2pz} is zero at the origin, positive above and negative below the xy plane, and zero at $z = \pm \infty$, it is a logical consequence that the $2pz$ atomic orbital must reach a maximum somewhere along the z axis above the origin and a minimum somewhere below.†

(d)   Finally, let us consider how χ_{2pz} varies in a horizontal plane parallel to xy and m units above the plane (cf. Fig. 1.1-D). The atomic orbital

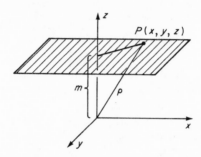

F IG. 1.1-D

function then is given by $(k^{5/2}m/(\pi)^{1/2})e^{-k\rho}$. Inspection of Fig. 1.1-D shows that as points in this plane increasingly distant from the z axis are considered, ρ increases and thus the negative exponential term decreases. Thus χ_{2pz} diminishes in horizontal directions away from the z axis.

From the prior four observations it is apparent that χ_{2pz} is schematically depicted by Fig. 1.1-E. In a similar manner, χ_{2px} is found to lie along the x axis and χ_{2py} along the y axis, each with a positive and negative lobe.

* The product $ze^{-k\rho}$ is said to be "indeterminate" as z approaches plus or minus infinity. The simplest question is the limiting value of this product in proceeding along the z axis to plus or minus infinity; here $\rho = z$.

$$\lim_{z \to \pm\infty} ze^{-kz} = \lim_{z \to \pm\infty} \frac{z}{e^{kz}} = \lim_{z \to \pm\infty} \frac{dz/dz}{d(e^{kz})/dz} = \lim_{z \to \pm\infty} \frac{1}{ke^{kz}} = 0$$

The first equality derives from the L'Hospital rule that the limit of an indeterminant quotient is given by the limit of the quotient of numerator and denominator derivatives.

A similar proof can be used to show that χ_{2pz} vanishes as one proceeds in *any* direction from the nucleus, that is, as $\rho \to \infty$.

† By considering only points along the z axis, that is, setting $\rho = z$, differentiating χ_{2pz} (cf. Eq. 1.1-1), and setting the derivative equal to zero, the reader can quickly demonstrate that these extrema occur at $z = \pm(1/k)$.

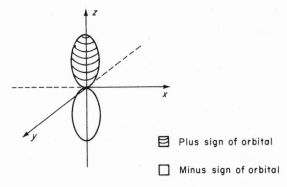

Fɪɢ. 1.1-E

1.1b   Electron Density and Normalization

More related to observational reality than an orbital itself is the square of the orbital, for it is the squared orbital function* which gives *electron density*. Thus, for example, the value of $\chi_{2p_z}{}^2 = (k^5/\pi)z^2e^{-2k\rho}$ at any point in space will give the fraction of an electron in this orbital per unit volume at this point.

As a result of this definition, an integration of an orbital squared over all space—that is, from $x = -\infty$ to $+\infty$, $y = -\infty$ to $+\infty$, and $z = -\infty$ to $+\infty$—must afford a value of one, for it is adding the electron density at all points in space. For example,

$$\int_{x=-\infty}^{+\infty}\int_{y=-\infty}^{\infty}\int_{z=-\infty}^{\infty}\chi_{2s}{}^2\,dx\,dy\,dz = 1, \qquad \int_{-\infty}^{\infty}\int_{-\infty}^{\infty}\int_{-\infty}^{\infty}\chi_{2p_y}{}^2\,dx\,dy\,dz = 1$$

$$\int_{-\infty}^{\infty}\int_{-\infty}^{\infty}\int_{-\infty}^{\infty}\chi_{2p_x}{}^2\,dx\,dy\,dz = 1, \qquad \int_{-\infty}^{\infty}\int_{-\infty}^{\infty}\int_{-\infty}^{\infty}\chi_{2p_z}{}^2\,dx\,dy\,dz = 1$$

$$1.1\text{-}2$$

Actual integration of the squares of 2s and 2p orbitals is given as a problem for the reader to do. This is seen in each case to give unity. This would not be so, if it were not for the coefficients $k^{5/2}/(3\pi)^{1/2}$ and $k^{5/2}/\pi^{1/2}$ occurring in the 2s and 2p orbitals of expression 1.1-1. These coefficients are called *normalizing* constants, and the process of multiplying an orbital by the

---

* A more involved statement would be required if complex orbital expressions were being considered.

proper constant to make the total electron density over all space equal to one is termed *normalization*.

1.1c Molecular Orbitals as Linear Combinations of Atomic Orbitals (LCAO)

When atomic orbitals overlap, the ensuing interaction results in formation of molecular orbitals. To the organic chemist, π systems are of particular importance; here the molecular orbitals result from interaction of all of the atomic p orbitals not involved in sigma bonding and generally parallel to one another. Mathematically, the molecular orbitals are taken as linear combinations of the component atomic orbitals (i.e., LCAO-MO). That is, the χ's (χ_1 from atom 1, χ_2 from atom 2, etc.) are added—however, not necessarily with equal weighting. Any given molecular orbital ψ, having n interacting atomic orbitals, thus may be written as

$$\psi = c_1\chi_1 + c_2\chi_2 + c_3\chi_3 + \cdots + c_n\chi_n \qquad 1.1\text{-}3$$

The c's are the weighting constants or *molecular orbital coefficients* indicating the extent to which each atomic orbital is weighted in the molecular orbital.

The LCAO-MO method can be termed a "mixing" process, for one does mix together atomic orbitals to obtain molecular orbitals. Several aspects of this process are presently noteworthy. First, when one quantum mechanically mixes a given number of atomic orbitals, more than one molecular orbital results. Each of the molecular orbitals obtained (ψ_1, ψ_2, ψ_3,...) has its own set of molecular orbital coefficients; that is, the weighting of the component atomic orbitals will differ in different molecular orbitals. Second, each molecular orbital has its own characteristic energy. Third, there will be as many molecular orbitals resulting from the mixing process as there were atomic orbitals mixed; in quantum mechanics orbitals are neither lost nor gained.

Let us consider the specific case of ethylene first. Here there are two atomic p orbitals to be mixed, χ_1 at atom 1 and χ_2 at atom 2. The method of mixing atomic orbitals is detailed subsequently; it affords both the energy of each molecular orbital resulting from mixing and also the LCAO form (i.e., the MO coefficients). However, for the present, the results of mixing of χ_1 and χ_2 will merely be given, for there are advantages to describing molecular orbitals further prior to detailing methods for obtaining them.

The two molecular orbitals resulting from interaction and quantum mechanical mixing of χ_1 and χ_2 are

$$\psi_1 = \frac{1}{\sqrt{2}}\,\chi_1 + \frac{1}{\sqrt{2}}\,\chi_2, \qquad \psi_2 = \frac{1}{\sqrt{2}}\,\chi_1 - \frac{1}{\sqrt{2}}\,\chi_2 \qquad \text{1.1-4a}$$

Since χ_1 and χ_2 are simple functions of coordinates, it is apparent that the two ethylenic molecular orbitals ψ_1 and ψ_2 in turn can be expressed as analytic functions of x, y, and z; that is, at every point P in space having a given set of coordinates, ψ_1 and ψ_2 will each have a definite value. Let us then ascertain the geometric properties of ψ_1 and ψ_2. If atomic orbitals χ_1 and χ_2 are taken centering at O_1 and O_2 and separated by the ethylenic interatomic distance R_{12} (cf. Fig. 1.1-F), then these are given by

$$\chi_1 = \frac{k^{5/2}}{\pi^{1/2}}\,z\,\exp(-k\rho_1), \qquad \chi_2 = \frac{k^{5/2}}{\pi^{1/2}}\,z\,\exp(-k\rho_2) \qquad \text{1.1-5}$$

and ψ_1 and ψ_2, as given by 1.1-4a, may be rewritten*

$$\psi_1 = \frac{k^{5/2}}{(2\pi)^{1/2}}\,z[\exp(-k\rho_1) + \exp(-k\rho_2)]$$

$$\psi_2 = \frac{k^{5/2}}{(2\pi)^{1/2}}\,z[\exp(-k\rho_1) - \exp(-k\rho_2)] \qquad \text{1.1-4b}$$

One could evaluate ψ_1 at several points in space. However, inspection of ψ_1 as given either in Eq. 1.1-4a or 1.1-4b indicates that it is the superposition of χ_1, centering at O_1 and χ_2 centering at O_2. Thus knowing the general geometric properties of atomic p_z orbitals, as discussed earlier we can begin by concluding that the superposition is reasonably depicted by Fig. 1.1-G. This is the ethylenic π orbital depicted in most elementary textbooks and familiar to most organic chemists. Several aspects are easily derived from the mathematical formulation of ψ_1 as given in Eq. 1.1-5. First, for points in the XY plane, where $z = 0$, ψ_1 is zero and has a node. Second, since

---

* Using the formula, $d_{ab} = [(x_a - x_b)^2 + (y_a - y_b)^2 + (z_a - z_b)^2]^{1/2}$, for the distance between two points in space $A(x_a, y_a, z_a)$ and $B(x_b, y_b, z_b)$, one could reformulate Eqs. 1.1-5 explicitly in terms of the variables x, y, and z alone; i.e.,

$$\psi_1 = \frac{k^{5/2}z}{(2\pi)^{1/2}}\,(\exp\{-k[(x + l)^2 + y^2 + z^2]^{1/2}\} + \exp\{-k[(x - l)^2 + y^2 + z^2]^{1/2}\})$$

$$\psi_2 = \frac{k^{5/2}z}{(2\pi)^{1/2}}\,(\exp\{-k[(x + l)^2 + y^2 + z^2]^{1/2}\} - \exp\{-k[(x - l)^2 + y^2 + z^2]^{1/2}\})$$

where $2l = R_{12}$, the interatomic distance.

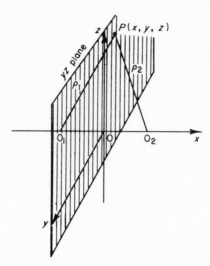

FIG. 1.1-F

exponentials are always positive, the sign of ψ_1 is determined by z; and ψ_1 is positive above the XY axis where Z is positive and negative below the XY plane. As with the components χ_1 and χ_2, ψ_1 approaches zero as ρ_1 and ρ_2, the distances from the nuclei, increase toward infinity. ψ_1 must then be a maximum somewhere between $z = 0$ and ∞. For points on any circle symmetrical about the X axis, ρ_1 and ρ_2 are constant, and ψ_1 is maximized in the XZ plane since Z is then maximized. Now looking at ψ_2, we see that this normalized difference between χ_1 and χ_2 is equivalent to the superposition of χ_1 and an inverted χ_2. This is clear once one realizes that taking χ_2 with a negative sign in the linear combination is equivalent

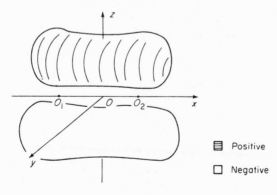

FIG. 1.1-G

to replacing z by $-z$. Thus $(-\chi_2) = (k^{5/2}/\pi^{1/2})(-z)\exp(-k\rho_2)$ *is* an inverted χ_2, for whenever z is positive $(-\chi_2)$ is negative and vice versa.

There is one salient feature of ψ_2, resulting from the equivalence but opposite signs of χ_1 and $-\chi_2$; this is the cancellation of these orbitals in the YZ plane. This is easily seen from the expression for ψ_2 as given in Eq. 1.1-5 once it is realized (cf. Fig. 1.1-F) that in the YZ plane $\rho_1 = \rho_2$ and thence $\psi_2 = 0$. Thus ψ_2 has a node not only in the XY plane but also in the YZ plane. Note Fig. 1.1-H describing ψ_2.

While ψ_1 is of lower energy than either of the component p orbitals, ψ_2

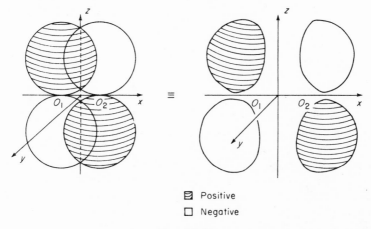

⊟ Positive
☐ Negative

Fig. 1.1-H

is higher. A molecular orbital whose energy is lower than a single p orbital is said to be *bonding* while a molecular orbital of higher energy than a single and isolated p orbital is termed *antibonding*. A molecular orbital whose energy is the same as that of an isolated p orbital is *nonbonding*. In the ethylenic case ψ_1 is bonding and ψ_2 is antibonding; taking the energy of noninteracting p orbitals as the arbitrary reference zero of energy, one can formulate the interaction of ethylenic p orbitals schematically and energy-wise (Fig. 1.1-I).

Two final aspects are to be noted in connection with the ethylene situation. First, since the π system of ethylene contains two electrons which will fill the low-energy molecular orbital, the π *energy* of ethylene may be said to be $-2|\beta|$ (i.e., $-1|\beta|$ per electron).* Clearly this is energetically better than having two "insulated" p orbitals containing two electrons, in which case the energy is zero. Second, in its simplest (Hückel) form the LCAO

---

* Pending a discussion of β, we can use the ethylenic case to define the energy units $|\beta|$ of Fig. 1.1-I as half the π bonding energy in ethylene.

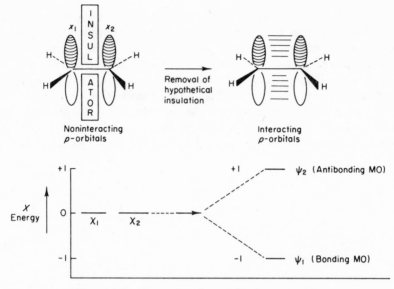

F<small>IG.</small> 1.1-I.   Formation of two MOs from two AOs.

theory gives the electron density per electron in a given molecular orbital
and at a given atom (i) by the square of the LCAO-MO coefficient (c_i^2).
Thus in the bonding MO of ethylene (ψ_1) $c_1^2 = (1/\sqrt{2})^2 = \frac{1}{2}$ and also
$c_2^2 = \frac{1}{2}$ give the π-electron density at atoms 1 and 2 for one electron. Since
electron density contributions are additive and ethylene has two electrons
in ψ_1, the π-electron density at atom 1 is $q_1 = 1$ and at atom 2 is $q_2 = 1$.
The even electron distribution is no great surprise.

1.1d *Summary of Salient Features of LCAO-MO Mixing*

(a) Mixing or quantum mechanical interaction of n atomic p orbitals
(the χ's) gives n π molecular orbitals, some of higher energy (antibonding),
some of lower energy (bonding), and sometimes some of the same energy
(nonbonding) compared to the atomic orbitals mixed.

(b) Orbitals in general and molecular orbitals specifically can accommodate at most two electrons per orbital. The lowest energy orbitals are
preferentially populated.

(c) The π energy of the system is the sum of the energies of all the π
electrons. The energy of an electron is that of the orbital it occupies.

(d) Each molecular orbital has its own LCAO form of the type $\psi =
c_1\chi_1 + c_2\chi_2 \cdots c_n\chi_n$, where the c's indicate the extent to which each
atomic orbital is weighted in the MO.

(e)   The π-electron density contributed by an electron to a given atom is given by the square of the coefficient of that atom in the MO containing the electron.

(f)   The total π-electron density at a given atom is the sum of the π-electron density contributions, as outlined in (e), due to all electrons.

(g)   Since the sum of the squares of the coefficients in any particular MO gives the total electron density for one electron in that MO and since this total must be one electron, the sum of squares of coefficients of a MO should equal unity. Such an orbital is said to be normalized.

1.2   Basic Procedure for Quantum Mechanical Mixing of Atomic Orbitals; Solution for Molecular Orbital Energies

1.2a   The Secular Determinant and Its Solution for Molecular Orbital Energies

In the preceding section the results of mixing two parallel p orbitals to give the ethylenic molecular orbitals were given without justification; this had the advantage of giving the reader a feeling for the phenomenon of orbital mixing prior to his actually learning how the mixing is done. The procedure followed in mixing atomic orbitals is now detailed; however, the theoretical justification is postponed until the reader has acquired a familiarity with the language and practicalities of quantum mechanical orbital mixing.

The general rules for mixing any set of atomic orbitals are given first and then applied to specific examples.

Rules for Quantum Mechanical Mixing

1.   Write a determinant having as many columns and as many rows as atomic orbitals to be mixed. The properties of determinants will be given as needed; for the present, a determinant can be considered to be a square array of numbers and variables evaluated in a specific way to be shown.

2.   Label the columns successively with the atomic orbitals to be mixed. That is, above column 1 place χ_1; above column 2, place χ_2, and so on. Then in the same order label the rows of the determinant.

3.   Beginning with the upper left element, where column 1 intersects row 1, and proceeding diagonally to the bottom right element, fill in X's along the diagonal. Each element of a determinant can be labeled in general a_{rs} where r refers to the row of the element and s refers to the column, and this rule sets $a_{rs} = X$ when $r = s$.

4.   Where a row corresponding to a given atomic orbital (χ_r) intersects

a column headed by another atomic orbital (χ_s), we have element a_{rs} which is a measure of the overlap of these two atomic orbitals. The number a_{rs} to be filled in is zero if the two atomic orbitals χ_r and χ_s are distant and noninteracting, while it is one if the orbitals are adjacent and overlapping. Although it is clear that there will be intermediate and varying extents of atomic orbital overlap, for the present we will consider only the extreme approximation where all nonvicinal orbital interactions are taken as zero.

5. Set the determinant equal to zero.

6. Solve the determinant for X. Each value of X obtained is a molecular orbital energy. The determinant described above is called the *secular determinant* and the equation in which the determinant is set equal to zero is the *secular determinantal equation*.

1.2b Application of LCAO Mixing to Ethylene

The simplest case to which the rules just given for mixing atomic orbitals can be applied is that of ethylene. Here χ_1 and χ_2, atomic p orbitals centering at atoms 1 and 2, are to be mixed. Since two atomic orbitals are being mixed, the determinant will have two rows and two columns (rule 1). Columns 1 and 2 are labeled with χ_1 and χ_2, and rows 1 and 2 are labeled in the same way (rule 2). The main diagonal elements (upper left to lower right) are filled in with X's (rule 3). The only nondiagonal elements in the ethylene example are a_{12} and a_{21}; both of these are filled in as 1, since these elements represent the interaction between χ_1 and χ_2 which are adjacent and overlapping atomic orbitals (rule 4). There are no nonoverlapping and noninteracting atomic orbitals and therefore no zero elements to be filled in. Finally, the resulting secular determinant is set equal to zero (rule 5), to give

$$\begin{array}{c} \\ \chi_1 \\ \\ \chi_2 \end{array} \begin{array}{c} \chi_1 \quad \chi_2 \\ \begin{vmatrix} X & 1 \\ 1 & X \end{vmatrix} \end{array} = 0 \qquad\qquad 1.2\text{-1a}$$

Two-by-two, or second-order, determinants are simply evaluated by taking the product of elements along the main diagonal, indicated by the dashed arrow below, and subtracting the product of elements along the alternative diagonal, indicated by the dotted arrow below*:

---

* This mnemonic device for evaluation of a 2×2 determinant derives from the general definition of determinants. A determinant, written as a square array of numbers and variables, is evaluated by definition as the sum of all possible products of the type $\pm a_{r1}a_{s2}a_{t3}$, etc., that is, products obtained by selecting an element from each column of the determinant—however, with the proviso that in this selection of elements for

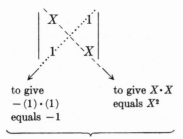

to give   to give $X \cdot X$
$-(1) \cdot (1)$ equals X^2
equals -1

Thus value of determinant is $X^2 - 1$

Hence on expansion, or evaluation, of the secular determinant, the secular equation 1.2-1a becomes

$$X^2 - 1 = 0 \quad \text{or} \quad X^2 = 1 \quad \text{and} \quad X = \pm 1 \qquad \text{1.2-1b}$$

$X = -1$ gives the energy of the (low-energy) bonding MO of ethylene. $X = +1$ gives the energy of the (high-energy) antibonding MO of ethylene. Energy is expressed in the positive unit $|\beta|$, "the absolute value of *beta.*" It is to be remembered that the energies are relative to the energy of an isolated p orbital whose energy is therefore taken as zero. Thus solution of the ethylenic secular determinantal equation 1.2-1 has resulted in precisely the molecular orbital energies given earlier without justification (cf. Fig. 1.1-I). One molecular orbital ($X = -1$) of lower energy and one molecular orbital ($X = +1$) of higher energy than either of the two atomic p orbitals mixed ($X = 0$) have resulted.

Although ethylene is indeed the simplest example for illustrating atomic

---

any given product the same row must not be used more than once. Additionally, each product $a_{r1}a_{s2}a_{t3}$, etc., is given a plus or minus sign depending on the "evenness or oddness" of the number of permutations required to convert the term into the zeroth permutation $a_{11}a_{22}a_{33}$ .... Thus we keep the second subscripts in order and exchange only the first subscripts as many times as needed to give the zeroth permutation. For example, in a 3×3 determinant, one of the products obtained by selecting an element from each column is $a_{31}a_{22}a_{13}$. This is related to the zeroth permutation $a_{11}a_{22}a_{33}$ by a single exchange of the first and third elements, however keeping the second subscripts in order and moving only the first subscript with the term. With a single permutation (i.e., an odd number of permutations), this term is given a negative sign. Had two permutations been needed, it would have been given a positive sign. A 2×2 determinant (i.e., Eq. 1.2-1) can be written generally as

$$\begin{vmatrix} a_{11} & a_{12} \\ a_{21} & a_{22} \end{vmatrix}$$

There are two possible products, each containing an element from one of the two columns; these products are $a_{11}a_{22}$ and $a_{21}a_{12}$. The first has no inversions of order and the second has one. Hence the determinant has the value $a_{11}a_{22} - a_{21}a_{12}$, fitting the mnemonic device given above.

orbital mixing, there is the simpler case of a single, isolated atomic orbital; this does not involve mixing. The secular determinant set up for (e.g.) χ_r as in Eq. 1.2-2 is 1×1 with a single element corresponding to the intersection of row χ_r with column χ_r and therefore taken as X. Using the definition of a determinant (footnote, p. 12), we see that in a 1×1 determinant there is only one possible "product" of elements, namely the single element itself. A 1×1 determinant is hence equal to the element. In the present instance this gives the energy X of the isolated p orbital χ_r as zero in accord with the earlier statement

$$\begin{array}{c} \chi_r \\ \chi_r |X| = 0 \end{array} \qquad \text{or} \qquad X = 0 \qquad\qquad 1.2\text{-}2$$

that the energy of an isolated p orbital would be taken as our reference point and as zero.

1.2c Application to the Allyl Species and Determination of Its Molecular Orbital Energies

The next level of difficulty involves molecules having 3×3 secular determinants; there are two of these, the allyl and the cyclopropenyl species. The allyl species is considered first.

By the allyl species is meant the linear chain of trigonal carbon atoms

$$\overset{1}{\text{CH}_2}\!\!-\!\!\overset{2}{\text{CH}}\!\!-\!\!\overset{3}{\text{CH}_2}$$

having parallel p orbitals: χ_1 at carbon-1, χ_2 at carbon-2, and χ_3 at carbon-3. Note Fig. 1.2-A. The ordinary C—C and C—H single, or sigma bonds, are assumed in the present approximation to be composed of relatively nonmobile electrons which do not have to be included in the calculation. The LCAO calculation mixes χ_1, χ_2, and χ_3 and determines the molecular orbitals (i.e., here, their energies) available to the molecule. Only once the

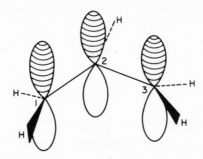

Fig. 1.2-A

MO energies are found and it is desired to total the π energy, is it important to know which allyl species is being considered and relatedly how many π electrons are to be accommodated in the MOs. If there are two π electrons, we are dealing with the allyl carbonium ion; if three, the allyl radical; if four, then the allyl carbanion. However, the secular determinant and its solutions are independent, in the present approximation, of the number of π electrons; the determinant and its solutions are functions only of the geometric ordering of the component p orbitals in space.

The secular equation for the allyl species is written by inspection (Eq. 1.2-3) using the rules of Section 1.2a:

$$
\begin{array}{c}
\begin{array}{ccc} \chi_1 & \chi_2 & \chi_3 \end{array} \\
\begin{array}{c} \chi_1 \\ \chi_2 \\ \chi_3 \end{array}
\begin{vmatrix} X & 1 & 0 \\ 1 & X & 1 \\ 0 & 1 & X \end{vmatrix} = 0
\end{array}
\qquad 1.2\text{-}3
$$

The three atomic orbitals χ_1, χ_2, and χ_3 label the rows and columns as required by rule 2; the ordering must be the same for the rows as for columns. According to rule 3, X's are written along the main diagonal (i.e., as elements a_{11}, a_{22}, a_{33}). Element a_{12} (i.e., row 1, column 2) derives from the interaction of χ_1 (row 1) and χ_2 (column 2). Since the atomic orbitals χ_1 and χ_2 are adjacent and overlapping, a one is filled in as this element in accord with rule 4. Similarly, element a_{21}, deriving from the interaction of the same two AOs, is filled in as a one. Elements a_{23} and a_{32} correspond to the interaction of adjacent and overlapping AOs χ_2 and χ_3 and are filled in as ones. Contrariwise, elements a_{13} and a_{31} are filled in as zeros since AOs χ_1 and χ_3 are not vicinal and are assumed not to interact. It now remains to evaluate the 3×3 secular determinant in Eq. 1.2-3 and then solve for X.

There is a simple mnemonic device for evaluation of third-order determinants somewhat similar to that given earlier for the second-order case. This can be demonstrated by its application to the secular determinant in 1.2-3. To solve such a third-order determinant, one repeats rows 1 and 2 in order below row 3. Through elements a_{11}, a_{21}, and a_{31} three dashed diagonal arrows are drawn as shown in the diagram below, each through three elements; the three elements lying on each dashed arrow are multiplied. Three triple products with a positive sign result. Now three dotted diagonal arrows are drawn through elements a_{13}, a_{23}, and a_{33}, each through three elements as shown. The three elements lying along each arrow are multiplied and the triple product obtained is given a minus sign; three such negative triple products result. The determinant then is the algebraic sum of the three positive (dashed) and the three negative (dotted) triple products*,†;

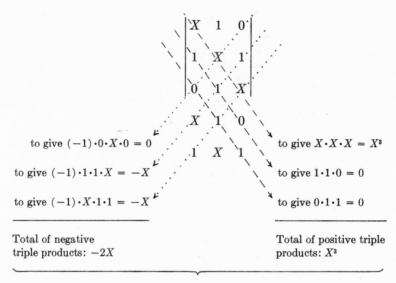

to give $(-1)\cdot0\cdot X\cdot0 = 0$

to give $(-1)\cdot1\cdot1\cdot X = -X$

to give $(-1)\cdot X\cdot1\cdot1 = -X$

to give $X\cdot X\cdot X = X^3$

to give $1\cdot1\cdot0 = 0$

to give $0\cdot1\cdot1 = 0$

Total of negative triple products: $-2X$

Total of positive triple products: X^3

Total value of determinant is $X^3 - 2X$

The secular determinantal equation 1.2-3 can now be rewritten as the cubic polynomial

$$X^3 - 2X = 0 \qquad \text{or} \qquad X(X^2 - 2) = 0 \qquad\qquad 1.2\text{-}4$$

Therefore, the solutions are $X = 0$ and $X = \pm\sqrt{2}$. Three atomic orbitals were mixed and three molecular orbitals have resulted. As in the ethylenic problem, one may envisage as in Fig. 1.2-B some imaginary insulators, capable of preventing interatomic orbital overlap, being removed. The interaction on removal of the insulator leads to splitting of the energy levels and formation of three molecular orbitals: a bonding MO at $-\sqrt{2}$, a nonbonding MO (i.e., at 0), and an antibonding MO at $+\sqrt{2}$ (cf. Fig.

* It may be seen that this mnemonic device merely provides an easy way to select all triple products of the type $a_{r1}a_{s2}a_{k3}$ as required by the definition of a determinant (cf. footnote p. 12) and also automatically determines whether the number of inversions of order is even or odd. All of the dashed products have zero or an even number of inversions of order while the dotted products have an odd number; this is why the dashed products are taken positively while the dotted ones are taken negatively.

† It is important for the reader to realize that the device used for second- and third-order determinants cannot be extended to higher order (e.g., 4 × 4) determinants. One can readily prove to himself that all (e.g.) quadruple products are not provided by such a scheme.

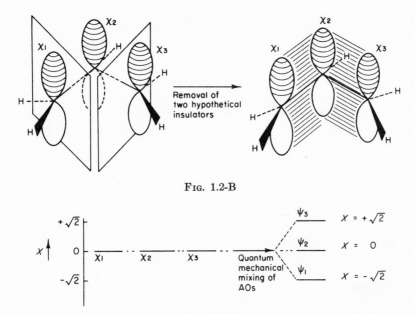

FIG. 1.2-B

FIG. 1.2-C

1.2-C). The three molecular orbitals formed are designated ψ_1, ψ_2, and ψ_3. For the present only the energy characteristics of the MOs will be considered; although each MO has its own spatial distribution as given by its LCAO-MO coefficients, the determination of coefficients will be delayed.

It is of interest now to determine the π energy of the allyl carbonium ion, the allyl radical, and the allyl carbanion. The allyl carbonium ion $CH_2\!\!=\!\!CH\!\!-\!\!CH_2^{\oplus}$ has two electrons. The allyl free radical $CH_2\!\!=\!\!CH\!\!-\!\!\dot{C}H_2$ has three, while the allyl carbanion $CH_2\!\!=\!\!CH\!\!-\!\!CH_2\!:^{\ominus}$ has four. Figure 1.2-D gives the MO energy diagrams with electrons assigned to the lowest possible orbitals, each orbital accommodating at most two electrons. The

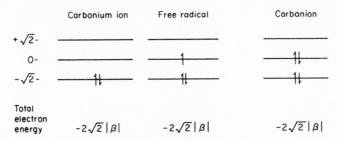

FIG. 1.2-D. Allyl carbonium ion, free radical, and carbanion π-electron energies.

π-electron energies, obtained by summing the π energies of the individual electrons, are given below each diagram. It is noted that all three species have the same π-electron energy of $-2\sqrt{2}|\beta|$; this is because the third and fourth electrons are introduced into the nonbonding MO and neither add to nor detract from the π energy of the system. The π energy of a system of carbon p orbitals can be pictured to be the energy gained on allowing the component p orbitals to interact; in the present instance this is the energy gained by removal of the hypothetical insulators as depicted in Fig. 1.2-B.

We have compared the allyl species with a hypothetical system in which the three atomic orbitals are completely noninteracting. Another model for comparison is both of present interest and of general significance. In this comparison we use a model corresponding to one of the contributing resonance structures of the system being considered. By determining the π energy of such a system "frozen" into one resonance contributor and comparing this with the π energy of the actual species in which complete overlap and delocalization is possible, we obtain the energy resulting from allowing resonance or delocalization. Thus the π-energy difference between the "frozen" and the delocalized species is termed the *delocalization, or resonance, energy.* In the present instance the "frozen" model has one ethylenic double bond and one isolated p orbital. Again one can picture removal of a hypothetical insulator, this time from between the ethylenic system and the single p orbital. Such a process leading to the fully delocalized allyl species is pictured in Fig. 1.2-E. In energy-level terminology we can write the process as in Fig. 1.2-F. Here we see that the starting frozen species prior to complete delocalization has the π systems of ethylene plus that of an isolated carbon p orbital. The former will contain two electrons,

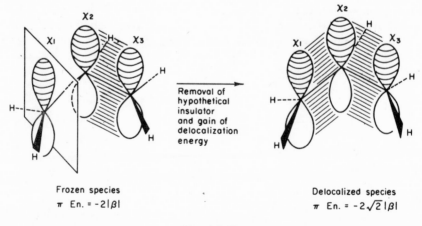

Frozen species
π En. = $-2|\beta|$

Removal of
hypothetical
insulator
and gain of
delocalization
energy

Delocalized species
π En. = $-2\sqrt{2}|\beta|$

Fig. 1.2.-E

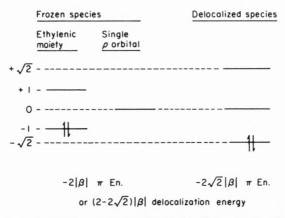

FIG. 1.2-F.   Electron delocalization in the allyl carbonium ion.

while in the case of the (e.g.) carbonium ion, the latter will not be occupied. After delocalization is allowed, the energy levels become those of the allyl species and two electrons fill the bonding MO. The increase in π energy accompanying the delocalization process, that is, the delocalization energy, is thus (cf. Fig. 1.2-F) $(2 - 2\sqrt{2})|\beta|$. One peculiarity of the allyl species, because of its nonbonding molecular orbital, is that the delocalization energy is the same for the carbonium ion, the free radical, and the carbanion. This may be seen in Fig. 1.2-F by adding the extra electron needed for the radical, and the extra two electrons in the case of the carbanion.

1.2d   Determination of the Molecular Orbital Energies for the Cyclopropenyl Species

The only other three-atom system of carbon p orbitals is cyclopropenyl (I), in which the three carbon p orbitals are arranged in a cyclic fashion.

(* = ⊕ for cyclopropenyl cation , • for the radical, ⊖ for the carbonion)

Since there are no noninteracting and nonoverlapping p orbitals the secular determinant in Eq. 1.2-5 has no zeros and only ones in the off-diagonal elements.

$$
\begin{array}{c}
\begin{array}{ccc} \chi_1 & \chi_2 & \chi_3 \end{array} \\
\begin{array}{c} \chi_1 \\ \chi_2 \\ \chi_3 \end{array}
\begin{vmatrix} X & 1 & 1 \\ 1 & X & 1 \\ 1 & 1 & X \end{vmatrix} = 0
\end{array}
\qquad 1.2\text{-}5
$$

Expanding this 3×3 secular determinant, using the mnemonic device given on page 16, one obtains

$$X^3 - 3X + 2 = 0$$

which factors into

$$(X + 2)(X - 1)(X - 1) = 0.$$

Thus the solutions for Eq. 1.2-5 are

$$X = -2, \quad X = +1, \quad \text{and} \quad X = +1$$

One could have obtained these energy levels from trial solution of the expansion of the determinant or by plotting $Y = X^3 - 3X + 2$ versus X and determining the values of X for which $Y = 0$.

Figure 1.2-G gives the π energies of the cyclopropenyl cation, free radical, and carbanion as well as the delocalization energies. The frozen model used for calculation of the delocalization energies (DEs) is that of one resonance contributor, that is, one ethylenic double bond plus one isolated p orbital having the balance of the electrons. This model has the π energy of $-2|\beta|$ for all three cyclopropenyl species, since the p orbital contributes zero regardless of the number of electrons assigned to it.

It is most interesting that the molecular orbital prediction is for greatest stabilization for the two-electron cationic species, less for the three-electron radical species, and least for the four-electron carbanionic species.* The simple resonance theory without added assumptions† would not have

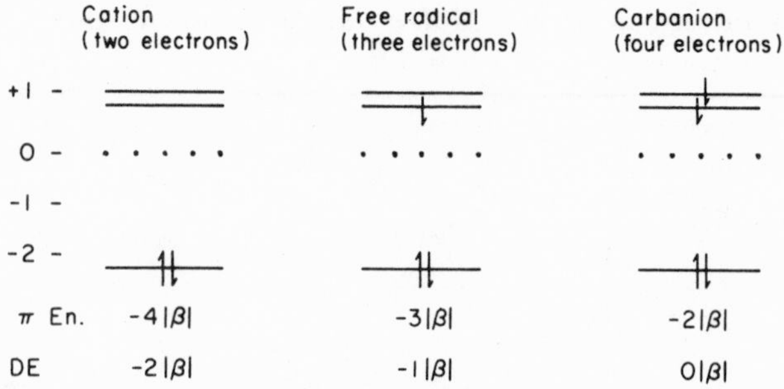

Fig. 1.2-G. Delocalization and π energies of the cyclopropenyl species.

* In the case of the cyclopropenyl carbanion there are two electrons available for the two levels at $+1$, and in the present approximation one would predict these two electrons to occupy different $+1$ levels with unpaired spins to give a triplet as a consequence of Hund's rule. An analogous situation obtains for the cyclobutadiene example discussed next.

† The Hückel rule (*vide infra*) provides such a differentiation, but this rule derives from molecular orbital theory.

provided this differentiation, since we have the same number of resonance contributors for all three species. Experimentally it is of course well known that the cationic species is most stable. Furthermore, this cyclic two-electron species constitutes the simplest example of the Hückel rule[4] specifying special stability for cyclic species having $4n + 2$ electrons.

1.2e Application of Direct Approach to Larger Molecules

Using the direct approach set forth for LCAO mixing thus far one can write the secular determinant for any π system composed of parallel p orbitals. However, expansion of even a fourth-order determinant requires a more involved treatment than encountered thus far. In many cases, especially where there is molecular symmetry, there are simpler approaches to the problem; and these are outlined shortly. Nevertheless, the present treatment is extended to two of the larger molecules. This allows introduction of cofactors which are of importance in the section to follow and also indicates the procedure available when simpler methods are not available.

The expansion of a determinant by cofactors is a useful procedure. The cofactor of any element a_{ij} in a determinant can be signified by A_{ij}. It is obtained by deleting that row (i.e., i) and that column (i.e., j) containing the element under consideration and then placing a plus sign in front of the resulting lower order determinant if the sum of i and j is even or a minus sign if the sum of i and j is odd. Thus in the fourth-order determinant D:

$$D = \begin{vmatrix} a_{11} & a_{12} & a_{13} & a_{14} \\ a_{21} & a_{22} & a_{23} & a_{24} \\ a_{31} & a_{32} & a_{33} & a_{34} \\ a_{41} & a_{42} & a_{43} & a_{44} \end{vmatrix}, \qquad A_{12} = - \begin{vmatrix} a_{21} & a_{23} & a_{24} \\ a_{31} & a_{33} & a_{34} \\ a_{41} & a_{43} & a_{44} \end{vmatrix} \qquad 1.2\text{-}6$$

the cofactor A_{12} of element a_{12} is obtained by deleting row 1 and column 2 to give a 3×3 determinant. A negative sign is prefixed since a_{12} is an odd element; that is, the subscripts add to give an odd number.

Any determinant of order n (i.e., an $n \times n$ determinant) may be expressed as a linear combination of $(n - 1)$-order determinants; this is termed expansion of a determinant by cofactors. One selects any row or column of the determinant and multiplies each element of that row or column by its cofactor. The original determinant is then given by the algebraic sum of these products of element and cofactors. For example, the determinant D of Eq. 1.2-6 could be expanded using the first row to give

$$D = a_{11}A_{11} + a_{12}A_{12} + a_{13}A_{13} + a_{14}A_{14} \qquad 1.2\text{-}7$$

Since each of the cofactors is a 3×3 determinant which we can expand by inspection by the device given on page 16, while the original fourth-order determinant is not subject to easy expansion, there is a clear advantage in expanding by cofactors. Actually one would select that row or column which had the maximum number of elements equal to zero, for this would give us the smallest number of 3×3 determinants (i.e., the cofactors) to expand. The same method can be used with a higher order determinant by successive application. Thus a fifth-order determinant can be expanded by cofactors into a linear combination of 4×4 determinants, each of which can then be expanded into 3×3's. Clearly, this method quickly reaches a practical limit.

Let us apply the method to both cyclobutadiene and methylenecyclopropene. In the case of cyclobutadiene (II) the four atomic p orbitals are labeled in order around the four-membered ring where the positive lobes

II

are pictured as projecting above the plane of the paper. The fourth-order secular determinant and the secular equation are then

$$
\begin{array}{c}
\begin{array}{cccc} \chi_1 & \chi_2 & \chi_3 & \chi_4 \end{array} \\
\begin{array}{c}\chi_1 \\ \chi_2 \\ \chi_3 \\ \chi_4\end{array}
\begin{vmatrix}
X & 1 & 0 & 1 \\
1 & X & 1 & 0 \\
0 & 1 & X & 1 \\
1 & 0 & 1 & X
\end{vmatrix} = 0
\end{array}
\qquad \text{1.2-8}
$$

Using the first row, we expand this by cofactors to obtain

$$
X\begin{vmatrix} X & 1 & 0 \\ 1 & X & 1 \\ 0 & 1 & X \end{vmatrix} + 1 \cdot (-1)\begin{vmatrix} 1 & 1 & 0 \\ 0 & X & 1 \\ 1 & 1 & X \end{vmatrix} + 1 \cdot (-1)\begin{vmatrix} 1 & X & 1 \\ 0 & 1 & X \\ 1 & 0 & 1 \end{vmatrix} = 0 \qquad \text{1.2-9}
$$

It is noted that there is no term involving the element a_{13} and cofactor A_{13} since the former is zero. Expansion of these determinants in 1.2-9 gives

$$
X(X^3 - 2X) - (X^2 + 1 - 1) - (1 + X^2 - 1) = X^4 - 4X^2
$$
$$
= X^2(X^2 - 4) = 0
$$

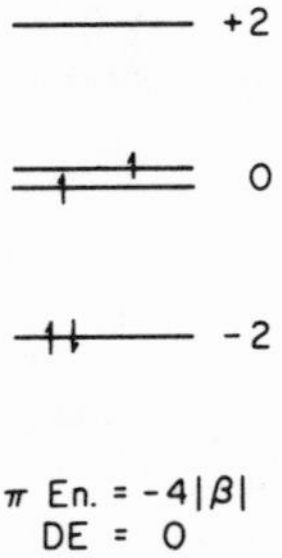

$$\pi \text{ En.} = -4|\beta|$$
$$DE = 0$$

FIG. 1.2-H. Cyclobutadiene MOs.

giving MO energy levels of $X = 0$, $X = 0$, $X = -2$, $X = +2$. Cyclobutadiene has four π electrons, and the energy levels may be filled and the π energy and π delocalization energy determined as in Fig. 1.2-H. The frozen model used for calculating delocalization energy here is one resonance contributor of cyclobutadiene in which the two double bonds are assumed not to interact; this model therefore has π energy which is double that of ethylene, or $-4|\beta|$. This is precisely the π energy of the delocalized cyclobutadiene, so that MO theory predicts no stabilization by delocalization. The organic chemist is well aware of the lack of stability of simple cyclobutadienes and we note its nonconformity with Hückel's $4n + 2$ electron requirement for aromaticity.

One further aspect is noteworthy. There are two energy levels at $X = 0$; these are said to constitute *a degenerate pair*. And, in general, the occurrence of more than one orbital of the same energy is termed *degeneracy*. With two electrons filling the bonding MO at -2, there are two electrons which are available for the degenerate pair of MOs. In such a case Hund's rule suggests that the two electrons will go one into each member of the degenerate MOs and that the two electrons will have the same spin. The reason why this configuration, having two unpaired electrons and termed a *triplet*, is of lower energy than the alternative configuration in which the electrons have opposite, or paired, spins is the subject of a later discussion. The present example is similar to that of the cyclopropenyl carbanion (cf. footnote on p. 20).

The case of methylenecyclopropene (III) is also instructive. Here the secular determinant may be expanded as before. However, the resulting polynomial can only be partially factored.

$$
D = \begin{array}{c} \\ \chi_1 \\ \\ \chi_2 \\ \\ \chi_3 \\ \\ \chi_4 \end{array} \begin{array}{cccc} \chi_1 & \chi_2 & \chi_3 & \chi_4 \\ \begin{vmatrix} X & 1 & 0 & 0 \\ 1 & X & 1 & 1 \\ 0 & 1 & X & 1 \\ 0 & 1 & 1 & X \end{vmatrix} \end{array} = X \cdot \begin{vmatrix} X & 1 & 1 \\ 1 & X & 1 \\ 1 & 1 & X \end{vmatrix} - 1 \cdot \begin{vmatrix} 1 & 1 & 1 \\ 0 & X & 1 \\ 0 & 1 & X \end{vmatrix} = 0
$$

$$D = X(X^3 + 2 - 3X) - (X^2 - 1) = X^4 - 4X^2 + 2X + 1$$

$$= (X - 1)(X^3 + X^2 - 3X - 1) = 0$$

Hence one energy level is given by $X = 1$. The other three MO energies are the roots of $Y = X^3 + X^2 - 3X - 1 = 0$. These are obtained by plotting Y versus X to determine for which values of X the function Y equals zero. These roots are found to be $X = -2.17$, -0.31, and $+1.48$. The energy levels, π En., and delocalization energy (DE) are given in Fig. 1.2-I.

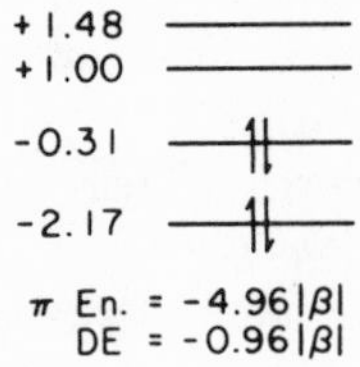

$$
\begin{array}{ll}
+1.48 & \text{———} \\
+1.00 & \text{———} \\
-0.31 & \text{—⥮—} \\
-2.17 & \text{—⥮—}
\end{array}
$$

$$\pi \text{ En.} = -4.96|\beta|$$
$$\text{DE} = -0.96|\beta|$$

Fig. 1.2-I.   Methylenecyclopropene MOs.

1.2f   Simple Mnemonic Device for Obtaining Molecular Orbital Energies for Unbranched Cyclic and Acyclic π Systems

A simple procedure has been described by Frost and Musulin[5] which allows one to write down quickly the MO energy levels for certain π systems. The case of simple, unbranched rings is considered first.

One begins by drawing a circle of radius $2|\beta|$. Then the appropriate regular polygon is inscribed in the circle; in doing this one vertex is placed at the bottom of the circle. If the molecular system is one of the cyclopropenyl species, an upside-down equilateral triangle (Fig. 1.2-J) is inscribed. If we are dealing with cyclobutadiene, we inscribe a square with one vertex down (Fig. 1.2-K). For any of the cyclopentadienyl species—carbanion, free radical, or cation—we draw in a pentagon (Fig. 1.2-L). For benzene,

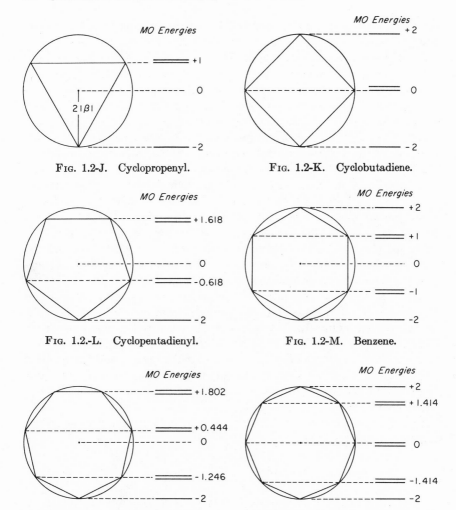

FIG. 1.2-J.   Cyclopropenyl.

FIG. 1.2-K.   Cyclobutadiene.

FIG. 1.2.-L.   Cyclopentadienyl.

FIG. 1.2-M.   Benzene.

FIG. 1.2-N.   Cycloheptatrienyl.

FIG. 1.2-O.   Cyclooctatetraene.

a hexagon (Fig. 1.2-M); for cycloheptatrienyl, a heptagon (Fig. 1.2-N); cyclooctatetraene, an octagon (Fig. 1.2-O); and so on.

Corresponding to each intersection of the polygon with the circle, there exists a MO whose energy is given by the vertical placement, that is, its projection on a vertical energy scale. The center of the circle is taken as the zero, the bottom as $-2|\beta|$, and the top as $+2|\beta|$. The vertical displacements from zero can be obtained by simple trigonometry; for many purposes a qualitative idea of the placement of the molecular orbitals is sufficient. For example, inspection of Figs. 1.2-J through 1.2-M reveals the source of the Hückel rule requiring $4n + 2$ electrons for aromaticity, where n is an integer. It may be seen that in all cases there is a single energy level at -2, thus requiring at least two electrons for a closed shell. Above the -2 level the MOs occur in degenerate pairs, requiring that all additional electrons be provided in (n) groups of four for each degenerate pair to give a closed-shell species. The summation of the electron requirement for closed-shell species is therefore $4n + 2$. Also, we note that species having $4n$ electrons are predicted in this approximation to be triplets, since two electrons are supplied for the highest occupied degenerate pair (cf. discussion on p. 23). Finally, since the low-energy species in each case will be those in which all of the bonding molecular orbitals are filled, making the π energy as negative as possible, inspection of these diagrams immediately suggests in each case which species should be heavily stabilized. From Fig. 1.2-J we correctly predict stability for the cyclopropenyl cation (two electrons); from Fig. 1.2-K we predict the cyclobutadienyl dication (two electrons) and the cyclobutadienyl dianion (six electrons) to be preferred. Figure 1.2-L suggests that the cyclopentadienyl carbanion (six electrons) should be favored over the radical and the cation in agreement with organic knowledge; similarly, Fig. 1.2-M leads us to the preference for neutral benzene in the six-ring system. Figure 1.2-N predicts special stability for tropylium cation (six electrons) compared to the anion and radical; the organic chemist is aware of the correctness here, too. Figure 1.2-O suggests that the dication (six electrons) and the dianion (10 electrons) will be the favored eight-ring aromatic species.* Thus, the Frost–Hückel circle mnemonic is, indeed, a convenient device.

An extension given by Frost and Musulin[5] allows application to unbranched acyclic π systems. For a chain of m atoms we draw (note Figs.

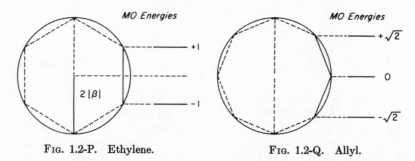

FIG. 1.2-P. Ethylene.   FIG. 1.2-Q. Allyl.

* In the past cyclooctatetraene has often been said to be nonaromatic because ring strain led to puckering. The argument is belied by the existence of planar cyclooctatetraene dianion where the π energy gained in attaining planarity is more than the strain energy resulting. Thus cyclooctatetraene may be said to be nonplanar for electronic rather than mere steric reasons.

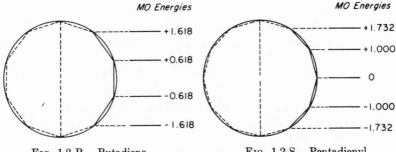

FIG. 1.2-R. Butadiene.   FIG. 1.2-S. Pentadienyl.

1.2-P–1.2-S) the polygon having $2m + 2$ sides inside the usual circle of radius $2|\beta|$ and with one vertex down as before. However, only those intersections of the polygon which are to the right of the vertical diameter of the circle are used. This excludes the intersections at the bottom of the circle (i.e., at $-2|\beta|$) and at the top of the circle (i.e., at $+2|\beta|$) as well as those to the left; the part of the polygon used corresponds to the linear π chain involved.

1.3 Elucidation of the Electronic Nature of Molecular Orbitals; Determination of LCAO-MO Coefficients

The preceding section concentrated on basic methods of eliciting molecular orbital energies. These were of value for determining the π and delocalization energies of numerous species of organic interest. This section deals with determining the electronic nature of these molecular orbitals. This is done by finding the LCAO-MO coefficients. With these coefficients in hand, giving the linear combination of atomic orbitals of which a particular molecular orbital is composed, we can obtain a mental picture of the molecular orbital as some given superposition of component AOs. Additionally, as will be shown, we can derive molecular properties as charge densities and bond order, and various reactivity indices as free valence.

1.3a A General Method of Determining LCAO-MO Coefficients

In the linear combination of atomic orbital–molecular orbital (LCAO-MO) method each molecular orbital is expressed in the form

$$\psi_j = c_{1j}\chi_1 + c_{2j}\chi_2 + c_{3j}\chi_3 + c_{4j}\chi_4 + \cdots + c_{nj}\chi_n \qquad 1.3\text{-}1$$

where the χ's are the atomic orbitals available to the π system being considered. This is similar to Eq. 1.1-3 except for the addition of subscript j which refers to MO j. This subscript is necessary, since for each different molecular orbital ψ_j there will be not only a characteristic energy (X_j) but also a different and characteristic set of LCAO coefficients, the c_j's. Therefore, for each different molecular orbital there will be a different admixture of atomic orbitals, as each c_{ij} indicates the weighting of atomic orbital χ_i in molecular orbital ψ_j. In labeling the coefficients it is customary to use the first subscript to refer to the atom and the second subscript to indicate the molecular orbital.

The same secular determinant which is used to derive the MO energies is now employed to ascertain the LCAO-MO coefficients. To do this, one selects any row or column; it is convenient to use row 1 routinely.* The elements of this first row of the secular determinant can be generally designated $a_{11}, a_{12}, a_{13}, \ldots, a_{1n}$ for an $n \times n$ determinant representing a molecule having n component atomic orbitals. The cofactors of the first-row elements can then parallel-wise be labeled $A_{11}, A_{12}, A_{13}, \ldots, A_{1n}$. It is now stated that these cofactors of the elements of the first row of the secular determinant give the unnormalized LCAO-MO coefficients $c_{1j}, c_{2j}, c_{3j}, \ldots,$ c_{nj}. That is, the cofactor of each element of the first row of the secular determinant gives the weighting constant for the orbital heading that column in which the element appears.

Thus if the secular determinant D is given by

$$D = \begin{array}{c} \\ \chi_1 \\ \\ \chi_2 \\ \\ \chi_3 \\ \cdot \\ \cdot \\ \cdot \\ \chi_n \end{array} \begin{array}{ccccc} \chi_1 & \chi_2 & \chi_3 & \cdots & \chi_n \\ \hline a_{11} & a_{12} & a_{13} & \cdots & a_{1n} \\ a_{21} & a_{22} & a_{23} & \cdots & a_{2n} \\ a_{31} & a_{32} & a_{33} & \cdots & a_{3n} \\ \cdot & \cdot & \cdot & & \cdot \\ \cdot & \cdot & \cdot & & \cdot \\ \cdot & \cdot & \cdot & & \cdot \\ a_{n1} & a_{n2} & a_{n3} & \cdots & a_{nn} \end{array}$$

then $c_{1j} = A_{11}$, $c_{2j} = A_{12}$, $c_{3j} = A_{13}$, etc.; or in general†

$$c_{ij} = A_{1i} \qquad\qquad 1.3\text{-}2$$

The cofactors are functions of X, the molecular orbital energy, and thus they give LCAO-MO coefficients which are functions of which energy level (X) is being considered. This is in agreement with the earlier statement that each energy level has its own set of coefficients.

---

* In a few situations it will prove to be advantageous to select some other row.

† A derivation of this relationship is given later.

The method of determining LCAO-MO coefficients is most readily learned from applications to molecules which have been considered from the energy viewpoint in Section 1.2. The simplest example is ethylene, whose MO has the general form

$$\psi_j = c_{1j}\chi_1 + c_{2j}\chi_2 \qquad\qquad 1.3\text{-}3$$

Depending on which MO is considered (i.e., whether $j = 1$ or 2), we will obtain one or the other of two sets of c_1 and c_2.

The first row of the ethylenic secular determinant is enclosed in a dotted rectangle to emphasize that we plan to take the cofactors of each element of this row and use these cofactors as unnormalized LCAO coefficients (cf. p. 21 for the method of obtaining cofactors*). It is to be

$$D = \begin{array}{c} \\ \chi_1 \\ \\ \chi_2 \end{array} \begin{array}{c} \chi_1 \quad \chi_2 \\ \begin{vmatrix} \overline{X} & 1 \\ 1 & X \end{vmatrix} \end{array} \qquad A_{11} = X, \qquad A_{12} = -1$$

noted that each atomic orbital's weighting (i.e., its coefficient) in the LCAO expression is given by the cofactor of the element immediately below that orbital. The cofactors, or unnormalized coefficients, thus obtained are listed in Table 1.3-1. As obtained by the method of cofactors

TABLE 1.3-1
UNNORMALIZED COEFFICIENTS FOR ETHYLENE

Cofactor	General value	Value for ψ_1 ($X = -1$)	Value for ψ_2 ($X = 1$)
A_{11} (giving unnormalized c_1)	X	-1	$+1$
A_{12} (giving unnormalized c_2)	-1	-1	-1

the relative value of the coefficients is seen to depend on X, that is, on which MO is being considered, the bonding MO with $X = -1$ or the antibonding MO with $X = +1$. The last two columns of this table give the unnormalized coefficients. The unnormalized MOs are given by

$$\psi_1 = -\chi_1 - \chi_2 \qquad \text{or} \qquad \psi_1 = \chi_1 + \chi_2 \qquad\qquad 1.3\text{-}4$$

and

$$\psi_2 = \chi_1 - \chi_2 \qquad\qquad 1.3\text{-}5$$

* Actually, in obtaining these cofactors, we get 1×1 determinants. But as noted previously, the general definition of a determinant as a summation of all permutations reveals that a 1×1 determinant is just the single number in the determinant.

The method of cofactors merely gives the *relative* values of the coefficients in any single MO, nothing more. Therefore the negative sign in the cofactors $A_{11} = -1$ and $A_{12} = -1$ (i.e., for ψ_1 and $X = -1$) is without significance. Multiplying through by -1 we obtain the second and more convenient form of Eq. 1.3-4.

The unnormalized MOs of Eqs. 1.3-4 and 1.3-5 are unsatisfactory. We have noted in Section 1.1b that the square of a coefficient gives the π-electron density q_r at that atom (r) due to one electron in the given molecular orbital (j) under consideration:

$$q_{rj} = c_{rj}{}^2 \qquad\qquad 1.3\text{-}6$$

A respectable molecular orbital will have a total electron density, summed over all atoms, of one electron for each electron put into that MO; that is, the sum of the squares of the coefficients in a MO should be one for the MO to be properly *normalized*.* The sum of the coefficients in Eq. 1.3-4 is $1^2 + 1^2 = 2$. Similarly, Eq. 1.3-5 gives $1^2 + (-1)^2 = 2$. These MOs have the peculiar property of giving a total electron density of 2 when containing one electron, thus the requirement for normalization. *The rule is to take the sum of the squares of the unnormalized coefficients and to divide each coefficient by the square root of this sum.* In this case we divide each unnormalized coefficient by $\sqrt{2}$ to obtain

$$\psi_1 = 1/(\sqrt{2})\chi_1 + 1/(\sqrt{2})\chi_2 \qquad\qquad 1.3\text{-}7$$

$$\psi_2 = 1/(\sqrt{2})\chi_1 - 1/(\sqrt{2})\chi_2 \qquad\qquad 1.3\text{-}8$$

Now the squares of the coefficients do add to one in each MO, and we note that these are the bonding and antibonding molecular orbitals of ethylene as given in Section 1.1. Also, the π-electron density at each atom of the ethylene molecule may be calculated. This is the sum of the individual contributions by each electron to that atom:

$$q_r = \sum_j n_j q_{rj} = \sum_j n_j c_{rj}{}^2 \qquad\qquad 1.3\text{-}9$$

Equation 1.3-9 generalizes this statement, giving the total π-electron density q_r at atom r of a molecule as the sum of the individual contributions (the q_{rj}'s) by each electron in MO j to atom r. Here n_j is the number of electrons in each MO j and may be termed the *occupation number*, and the summation is over all filled MOs. Applied to the case of ethylene, Eq. 1.3-9 gives $2 \cdot (1/\sqrt{2})^2 = 1$ π density at each atom. Since the ethylenic double bond contains two π electrons and ethylene is symmetrical, we shall have to await more complex molecules to observe values which could not intuitively have been predicted.

---

* This statement is true for the present Hückel method but will need modification in some future instances.

Since the present section deals with LCAO-MO coefficients, it is presently desirable to define another molecular property derivable from a knowledge of these coefficients. This is bond order. The contribution to the bond order between atoms r and s by an electron in MO j is given by Eq. 1.3-10 and has been termed[6] the partial bond order:

$$p_{rs,j} = c_{rj}c_{sj} \qquad\qquad 1.3\text{-}10$$

Just as the total electron density was given as the sum of the individual contributions, similarly the total bond order is given by such a summation:

$$p_{rs} = \sum_j n_j p_{rs,j} = \sum_j n_j c_{rj}c_{sj} \qquad\qquad 1.3\text{-}11$$

Here again n_j is the occupation number of MO j.

Applied to the ground state of ethylene in which there are two electrons in ψ_1 and where the contribution per electron is $(1/\sqrt{2})(1/\sqrt{2}) = \frac{1}{2}$, this affords a bond order of one. Again this is not surprising, for ethylene does have one localized π bond.

It is of some interest to determine the bond order of an electronically excited ethylene in which one electron has been promoted to ψ_2. The bond order contribution in ψ_2 is $(1/\sqrt{2})(-1/\sqrt{2}) = -\frac{1}{2}$, which cancels the $\frac{1}{2}$ bond order contributions from one electron in ψ_1 giving a zero total bond order. Thus in this excited state one would expect relatively free rotation due to the absence of π bonding and energy gain by parallel p-orbital overlap.

1.3b   Application to Determination of Coefficients of the Allyl Species

The secular determinant's first row is used again to determine the cofactors—A_{11}, A_{12}, A_{13}—of this row's elements. These cofactors are 2×2 determinants which are expanded to give respectively the weighting (unnormalized coefficients) of the orbitals heading the columns of the determinant.

$$D = \begin{array}{c} \\ \chi_1 \\ \chi_2 \\ \chi_3 \end{array} \begin{array}{ccc} \chi_1 & \chi_2 & \chi_3 \\ \boxed{\begin{array}{ccc} X & 1 & 0 \end{array}} \\ 1 & X & 1 \\ 0 & 1 & X \end{array}$$

$$A_{11} = \begin{vmatrix} X & 1 \\ 1 & X \end{vmatrix}, \qquad A_{12} = -\begin{vmatrix} 1 & 1 \\ 0 & X \end{vmatrix}, \qquad A_{13} = \begin{vmatrix} 1 & X \\ 0 & 1 \end{vmatrix}$$

Table 1.3-2 gives the values of the cofactors and unnormalized coefficients

TABLE 1.3-2

UNNORMALIZED LCAO COEFFICIENTS FOR ALLYL AND
NORMALIZING FACTORS

Cofactor	General value	For ψ_1 ($X = -\sqrt{2}$)	For ψ_2 ($X = 0$)	For ψ_3 ($X = +\sqrt{2}$)
A_{11}	$X^2 - 1$	1	-1	1
A_{12}	$-X$	$+\sqrt{2}$	0	$-\sqrt{2}$
A_{13}	1	1	1	1
Sum of squares of unnormalized coefficients		4	2	4
Normalizing factor		$\frac{1}{2}$	$1/\sqrt{2}$	$\frac{1}{2}$

for the three MOs ($X = -\sqrt{2}$, $X = 0$, $X = +\sqrt{2}$). Also, for each MO the sum of the squares of the unnormalized coefficients is given along with the reciprocal of the square root of the sum. The latter are the normalizing factors by which each unnormalized MO is multiplied to effect normalization. Using the data in Table 1.3-2, we obtain the unnormalized allyl MOs as

$$(X = +\sqrt{2}) \qquad \psi_3 = \chi_1 - \sqrt{2}\chi_2 + \chi_3$$

$$(X = 0) \qquad \psi_2 = \chi_1 - \chi_3 \qquad\qquad 1.3\text{-}12$$

$$(X = -\sqrt{2}) \qquad \psi_1 = \chi_1 + \sqrt{2}\chi_2 + \chi_3$$

Multiplying each by the appropriate normalization factor from Table 1.3-2 we can write the normalized allyl MOs:

$$(X = +\sqrt{2}) \qquad \psi_3 = \tfrac{1}{2}\chi_1 - (1/\sqrt{2})\chi_2 + \tfrac{1}{2}\chi_3$$

$$(X = 0) \qquad \psi_2 = (1/\sqrt{2})\chi_1 - (1/\sqrt{2})\chi_3 \qquad 1.3\text{-}13$$

$$(X = -\sqrt{2}) \qquad \psi_1 = \tfrac{1}{2}\chi_1 + (1/\sqrt{2})\chi_2 + \tfrac{1}{2}\chi_3$$

We note from the coefficients that an electron in ψ_1 distributes itself more heavily at the central carbon than at the ends of the molecule, whereas in ψ_2 electrons are localized at the two end carbon atoms. We recognize from our earlier discussion in Section 1.1 that the consequence of a change in sign of coefficients of adjacent AOs is the appearance of a node, or region of no electron density. Said differently, in proceeding along the top side of a π system, from atom to atom, we can be certain that somewhere in between two points where the wave function has changed signs, there is a point where the wave function and therefore the electron density goes through zero. In ψ_1 there is no such node; in ψ_2 there is a node at atom 2; and in

TABLE 1.3-3
π-Electron Densities and Bond Orders for the
Allyl Species

Species	Q_1	Q_2	Q_3	P_{12}	P_{23}	(P_{13})
Cation	0.5	1.0	0.5	$1/\sqrt{2}$	$1/\sqrt{2}$	$\frac{1}{2}$
Radical	1.0	1.0	1.0	$1/\sqrt{2}$	$1/\sqrt{2}$	0
Carbanion	1.5	1.0	1.5	$1/\sqrt{2}$	$1/\sqrt{2}$	$-\frac{1}{2}$

ψ_3 there are two nodes, one between atoms 1 and 2 and one between atoms 2 and 3.

The total π-electron densities and bond orders are given in Table 1.3-3 for the allyl cation (two π electrons), allyl radical (three electrons), and the allyl carbanion (four electrons). Realizing that a carbon atom having a π-electron density of one has a zero formal charge, we see that the cation has no formal charge at the central carbon atom but half a formal positive charge at the end carbon atoms. This is in agreement with the resonance prediction. The radical has unit π-electron density and a zero formal charge at all atoms. The carbanion has no charge at the central carbon atom but a half-negative formal charge at atoms 1 and 3, which again accords with the resonance theory. It is noteworthy that π bond orders between atoms 1 and 3 can be calculated although, strictly speaking, a bond order should not be calculated between two atoms which in the original calculations were assumed not to overlap. The results obtained are presently used to indicate the tendency toward bonding rather than pursued quantitatively since we are dealing here with a rough perturbation calculation. We find that for the allylic carbanion, atoms 1 and 3 are antibonding; that is, the total bond order is negative and overlap would lead to destabilization. This is reasonable, since complete 1–3 overlap would afford the cyclopropenyl anion which has no delocalization energy compared to $2 - \sqrt{2}$ for the allyl carbanion. The zero 1–3 bond order for the allyl radical, suggesting little change in π energy on forming a 1–3 bond, predicts approximately correctly, for there is little gained in forming a 1–3 bond and converting the allyl radical (π En. $-2.83|\beta|$) to the cyclopropenyl radical (π En. $-3.00|\beta|$). For the allyl carbonium ion the 1–3 bond order is $+\frac{1}{2}$ and we would predict 1–3 bonding and formation of the cyclopropenyl cation to be favorable, which is correct. The allyl carbonium ion has a π energy of only $-2\sqrt{2}|\beta|$, or $-2.83|\beta|$, compared to the more stable cyclopropenyl cation, of π energy $-4|\beta|$. This approach, informing us which added π bonds will lead to more stable species and which will lead to less stable species, is discussed from a more quantitative and rigorous viewpoint in a later discussion of first-order perturbation theory.

1.3c   Cyclopropenyl Coefficients; Difficulties Due to Degeneracy

Logically, we would next proceed to the cyclopropenyl system and determine the MO coefficients. However, as will be seen, this is only partially possible using the present level of sophistication. The secular determinant and cofactors of the first row are

$$D = \begin{array}{c} \\ \chi_1 \\ \chi_2 \\ \chi_3 \end{array} \begin{array}{ccc} \chi_1 & \chi_2 & \chi_3 \\ \begin{vmatrix} X & 1 & 1 \\ 1 & X & 1 \\ 1 & 1 & X \end{vmatrix} \end{array}$$

$$A_{11} = \begin{vmatrix} X & 1 \\ 1 & X \end{vmatrix}, \qquad A_{12} = - \begin{vmatrix} 1 & 1 \\ 1 & X \end{vmatrix}, \qquad A_{13} = \begin{vmatrix} 1 & X \\ 1 & 1 \end{vmatrix}$$

Table 1.3-4 gives the values of the cofactors for the different cyclopropenyl MOs (i.e., $X = -2$, $X = +1$, $X = +1$). This calculation reveals that for the bonding MO ($X = -2$) of cyclopropenyl the cofactors, and thus the LCAO coefficients, are all equal. When properly normalized these coefficients become $1/\sqrt{3}$ and the bonding MO is

$$\psi_1 = (1/\sqrt{3})\chi_1 + (1/\sqrt{3})\chi_2 + (1/\sqrt{3})\chi_3 \qquad \qquad 1.3\text{-}14$$

The π-electron density per electron in the bonding MO is $\frac{1}{3}$ and therefore is $\frac{2}{3}$ for the two electrons in the bonding MO of the cyclopropenyl cation (IV). The formal charge at each of the three carbon atoms is predicted to be $+\frac{1}{3}$ in accord with the resonance picture.

However, the coefficients for the antibonding, degenerate pair of MOs at $+1$ cannot be obtained from the 2×2 cofactors, since these cofactors

TABLE 1.3-4

COFACTORS DERIVED FROM CYCLOPROPENYL SECULAR
DETERMINANT

Cofactor	General value	For $X = -2$ (ψ_1)	For $X = +1$ (ψ_2 or ψ_3)
A_{11}	$X^2 - 1$	3	0
A_{12}	$-X + 1$	3	0
A_{13}	$1 - X$	3	0

(**IV**)

become zero for $X = +1$. Because the cofactors merely give us the relative magnitude of the LCAO coefficients, obtaining zero for all cofactors does not imply that the coefficients are zero. The ratio is said to be indeterminate, and the unnormalized coefficients cannot be obtained in this way. We shall return to determining these coefficients once we have reached the next level of sophistication.

For the present, it can be stated generally that the cofactors derived from any given secular determinant will necessarily be zero whenever these correspond to a degenerate pair of MOs. The basis of this rule is discussed in Chapter 2.

It can be predicted that due to degeneracy at $X = 0$ all of the cyclobutadiene coefficients will not be derivable from the cofactors of the fourth-order secular determinant. The coefficients for the butadiene problem could be determined by the methods presented thus far; but, at this stage of molecular complexity, the calculations are sufficiently laborious compared to simpler available methods that we postpone the discussion of butadiene and larger molecules until Chapter 3 as well as the introduction of methods of simplification by use of symmetry properties.

1.4 Choice of the Basis Set of Atomic Orbitals in LCAO-MO Calculations

In the preceding examples, the atomic $2p$ orbitals which were quantum mechanically "mixed" through use of the secular determinantal equation were taken as all oriented with the plus signs aimed in the same direction. The molecular orbital coefficients obtained then indicated how these atomic orbitals were to be oriented and weighted in each MO; a plus sign indicated that the atomic orbital was to be aimed upward while a negative sign specified that in the given MO the AO was to be aimed downward.

Now the original set of AOs prior to mixing is termed the "basis set" employed. This merely means that we have somewhat arbitrarily chosen a set of orbitals to use in the mixing process. Actually, a different set might have been chosen, although there are some restrictions imposed.* For example, another permissible set of basis orbitals would have one (e.g., χ_r) or more of the complete set of p orbitals inverted with the plus and minus signs exchanged. When using such a basis set, one must not forget that in any MO each atomic orbital symbol represents that AO as it occurs in the basis set. Thus, a positive coefficient for χ_r signifies an inverted AO at atom r. In setting up a secular determinant with some of the basis orbitals inverted, we put in a -1 for each adjacent set of orbitals in which the plus lobe of one orbital is near the minus lobe of the other orbital, and vice versa.

It is found that the molecular orbital energies afforded by quantum mechanical mixing are independent of the orientation of the basis set of atomic orbitals mixed. Conversely, the molecular coefficients will be a function of the choice of the basis set. However, it will be noted that, although the algebraic form of the MOs depends on the choice of the basis set, the actual molecular orbital will not depend on this choice. Thus the coefficients will merely reinvert those orbitals in a MO which were taken upside down in the basis set; that is, if the orbital χ_r is an inverted one, it is the negative of a more conventionally chosen orbital aimed upward, and in each MO obtained using the inverted orbital χ_r the coefficient will be the negative of that obtained in the conventional calculation. Thus the direction of the AOs in the molecular orbitals found will be independent of the definition of the basis set.

A simple example is found in the case of the allyl species if we choose the basis, or starting, set of AOs with the p orbital at carbon-3 inverted (Fig. 1.4-A). Here the secular determinant becomes

$$
\begin{array}{c}
\quad\;\; \chi_1 \quad \chi_2 \quad \chi_3 \\
\begin{array}{c} \chi_1 \\ \chi_2 \\ \chi_3 \end{array}
\begin{vmatrix} X & 1 & 0 \\ 1 & X & -1 \\ 0 & -1 & X \end{vmatrix} = X^3 - 2X = 0 \quad\text{and}\quad X = 0,\ \pm\sqrt{2} \qquad 1.4\text{-}1
\end{array}
$$

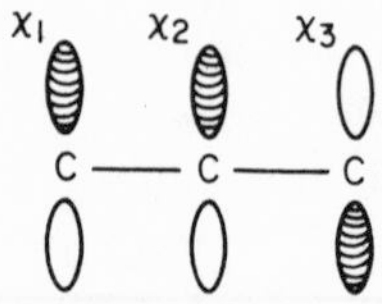

Fig. 1.4-A

* It is necessary that a "complete set" be chosen. For the present purpose this merely requires that all atomic orbitals be included.

TABLE 1.4-1

Cofactor (normalized coefficient[a])	General value	For $X = -\sqrt{2}$ (ψ_1)	For $X = 0$ (ψ_2)	For $X = +\sqrt{2}$ (ψ_3)
A_{11} (c_1)	$X^2 - 1$	1 $(\frac{1}{2})$	-1 $(1/\sqrt{2})$	1 $(\frac{1}{2})$
A_{12} (c_2)	$-X$	$\sqrt{2}$ $(1/\sqrt{2})$	0 (0)	$-\sqrt{2}$ $(-1/\sqrt{2})$
A_{13} (c_3)	-1	-1 $(-\frac{1}{2})$	-1 $(1/\sqrt{2})$	-1 $(-\frac{1}{2})$

[a] Normalized coefficients are given in parentheses.

The LCAO-MO coefficients are given in Table 1.4-1. Inspection of the coefficients of Table 1.4-1 shows that it gives the usual allyl coefficients with the exception that in every instance the sign of the coefficient c_3 is inverted. However, this is the coefficient weighting the present χ_3 which is the negative of the more common convention, and the coefficient serves to give an MO in each instance which has the AOs aimed with positive signs in the usual direction*.

1.5   Cases Where Negative Overlap Is Enforced in the Basis Set

Occasionally a molecular situation is encountered in which it is impossible to assign the direction of the basis set of AOs so that lobes of equal sign overlap. Thus "twist-hydrotrimethylenemethane" (I) might have its basis set of AOs chosen either as in Ia or in Ib, but in no case could one find a basis set with only plus lobes overlapping plus lobes and only minus lobes overlapping minus lobes. As we have recognized, either basis set will afford the same solution. However, for convenience the basis set Ib is

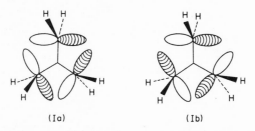

(Ia)   (Ib)

* There is one consequence of choosing a basis set with lobes of unequal sign overlapping. This is that bond order definitions given by Eqs. 1.3-10 and 1.3-11 must be modified to $p_{rs,j} = c_{rj}c_{sj}\epsilon_{rs}$ (1.3-10a) and $p_{rs} = \sum_j n_j c_{rj} c_{sj} \epsilon_{rs}$ (1.3-11a) where ϵ_{rs} is $+1$ or -1 depending on whether lobes of the same or opposite sign overlap.

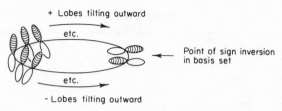

Fig. 1.5-A.   Möbius strip problem.

better since all of the off-diagonal elements will be the same, namely -1. As a problem (Problem 7 at the end of this chapter) it is suggested that the reader obtain the MO energies for "twist-hydrotrimethylenemethane"; these turn out to be $X = -1, -1$, and $+2$.

If one focuses attention on basis set Ia, he notes the presence of a cyclic array of contiguous p orbitals with like signs overlapping except for *one inversion*. This is related to the Möbius strip problem described by Heilbronner.[7] We may envisage a large cyclic polyene in which each p orbital is twisted a bit relative to the adjacent p orbital and finally the ends of this twisted chain are joined to give a cyclic polyene with a single inversion of p-orbital sign in proceeding from one atom to the next (Fig. 1.5-A). It has

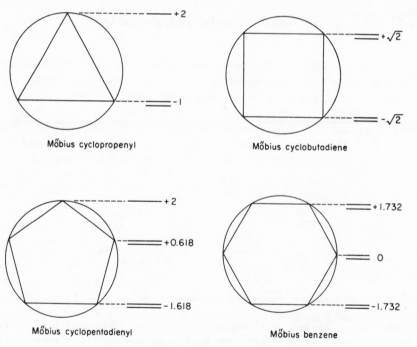

Fig. 1.5-B.   Möbius cyclic polyenes.

been noted by Heilbronner that in such a cyclic polyene the MO energy is given by the general expression*

$$X = 2 \cos \left[\frac{2L + 1}{n} \pi \right].$$

Here n is the number of AOs and L has the values $0, 1, \ldots, (n - 1)$ with each value of L giving one molecular orbital.

It is not difficult to demonstrate[8] that for such Möbius systems a simple mnemonic trick similar to the Frost–Hückel device is available. As with the Frost device, one inscribes the appropriate polygon in a circle of radius $2|\beta|$, and again the center of the circle is taken as the zero. As before, intersections of the polygon with the circle correspond to MO energy levels. However, in the case of Möbius strip cyclic polyenes, one orients one side of the polygon horizontally at the bottom of the circle (cf. Fig. 1.5-B).

1.6 The Hückel and Möbius Rules

Having developed the circle mnemonics for the Hückel and Möbius systems, we find it appropriate to use these in explaining the well-known Hückel rule which states that for a cyclic system of basis orbitals, $4n + 2$ confers particular stability. Also, we can inquire if this rule applies to Möbius systems, and if not, what rule is appropriate.

Looking at the circle mnemonic as applied to Hückel systems we note that invariably a single MO occurs at -2 and that this will accommodate two electrons. Above this in energy, the MOs occur in degenerate pairs (except for the highest occupied MO of even systems). Each pair will accommodate four electrons; if there are n degenerate pairs, $4n$ electrons are needed for these. This means that the total number of electrons required for a closed shell in a Hückel system will be $4n + 2$.

It is readily seen from the Möbius mnemonic[8] that a different rule is required since there is no single nondegenerate bonding MO. Since all the MOs (except for a highest energy MO in odd cases) come in degenerate pairs, the number of electrons needed for a stable, closed shell is $4n$. Thus for aromaticity, the Möbius rule is quite different.

In Hückel systems, a molecule having the wrong number of electrons for a closed shell has been defined[9] as "antiaromatic." The same definition

---

* Heilbronner's expression actually contained an extra factor to take into account the decreased overlap of noncoplanar p orbitals. For simplicity we merely use the decreased resonance integral as our energy unit.

TABLE 1.6-1

SUMMARY OF AROMATICITY AND ANTIAROMATICITY
REQUIREMENTS OF HÜCKEL AND MÖBIUS SYSTEMS

Type of system	Number of electrons	
	$4n + 2$	$4n$
Hückel	Aromatic	Antiaromatic
Möbius	Antiaromatic	Aromatic

could now be applied for Möbius systems, except that $4n + 2$ leads to antiaromaticity. Table 1.6-1 summarizes the situation.[8,10,11]

It has been noted[8] that not just twisted cyclic polyenes but any cyclic array of orbitals having an odd number of sign discontinuities constitutes a Möbius system. The Hückel systems similarly are more general. Also, both Zimmerman[8,11] and Dewar[10] have pointed out that the generalizations above apply not only to molecules but also to transition states.

1.7   Relation between the LCAO-MO Coefficients and the Molecular Orbital Energies

A very useful relation exists between the LCAO-MO coefficients of a given MO and the MO's energy. It will be shown later that minus twice the sum of the partial (one electron) bond orders, including all pairs of adjacent atoms, gives the energy of that particular MO in the usual units of $|\beta|$. We can picture the π bonding energy in any single MO as being dissected into contributions with the contribution from any given bond being given by the negative of twice the partial bond order

$$\Delta X_{rs,i} = -2p_{rs,i} \qquad\qquad 1.7\text{-}1$$

Thus a positive partial bond order between atoms r and s in MO i affords a negative contribution to the π energy (i.e., stabilization) while a negative bond order signifies antibonding.

We could, for example, take the LCAO-MO coefficients for allyl's lowest molecular orbital and obtain from these the MO energy. Thus the 1,2-partial bond order is (cf. Table 1.3-2 or 1.4-1 for coefficients) $1/2\sqrt{2}$; and the 2,3-partial bond order is the same. The total bond order in the bonding MO is therefore $1/\sqrt{2}$; and the π bonding energy in this MO is hence $-\sqrt{2}$ per electron. This is synonomous with the MO energy.

This energy–coefficient relationship has another use, that of deriving the

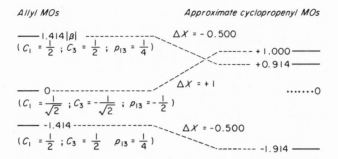

Fig. 1.7-A. Perturbation calculation of cyclopropenyl MO energy levels.

energy levels of one molecule from those of a structurally related one. For example, it is possible for each allyl MO to derive (Fig. 1.7-A) the energy change on allowing 1,3-bonding. We then obtain an approximation to the MO energies of cyclopropenyl. Since the original coefficients are derived from the allyl secular determinant which assumed no overlap between atoms 1 and 3, we cannot expect the energies obtained to be exact for cyclopropenyl. Precise Hückel energies could be derived, of course, using the cyclopropenyl coefficients.

Problems

1. Show that a Slater $2s$ orbital vanishes as $\rho \to$ infinity.
Hint: From the calculus of variations we know

$$\lim_{u \to \infty} \frac{f(u)}{g(u)} = \lim_{u \to \infty} \frac{f'(u)}{g'(u)}$$

and we note that χ_{2s} can be written in this quotient form.
2. In a p orbital, locate the point of maximum electron density.
3. Set up the secular determinant for benzene.
4. Obtain the Hückel molecular orbital energy levels for the species trimethylenemethane:

$$CH_2 = C \underset{\dot{C}H_2}{\overset{\dot{C}H_2}{<}}$$

5. Show that without methods beyond those given in Chapter 1 it is impossible to obtain all of the molecular orbital coefficients due to degeneracy in trimethylenemethane.

6. Convince yourself that a linear (i.e., acyclic) three-orbital system, in which AO 1 is perpendicular to AO 2, breaks down into a problem of two separate molecules. Which two? Set this problem up in a 3×3 determinant. What characteristic of the 3×3 determinant correlates with its ability to be "broken down" into two simpler problems?

7. Given "twist-hydrotrimethylenemethane," with three p orbitals in a

planar cyclic array, obtain the eigenvalues. How does this compare with the array of MO energies in cyclopropenyl itself? In each case decide which should be the stable species: the cation, the radical, or the carbanion?

8. Assume that the off-diagonal elements representing the interaction between AOs r and s are given by $\cos \Theta_{rs}$. Then obtain the MO energies for twisted ethylene as a function of angle of twist: for $0°$, $30°$, $60°$, $90°$, $120°$, $150°$, and $180°$. Put all this together then into a correlation diagram, plotting MO energies against Θ_{12}.

9. How does the number of MOs correspond to the number of basis AOs? Which of the following has the MOs symmetrically disposed about zero: Cyclopropenyl, ethylene, methyl, allyl? Can you correlate any molecular situation to the symmetrical disposition?

10. Derive the molecular orbital energies for butadiene. To do this use the method of expansion by cofactors of the fourth-order secular determinant.

11. Show that the molecular orbital energies obtained for the cyclopropenyl problem are the same despite a basis set of atomic orbitals being chosen having χ_3 inverted.

12. Derive the MO energy levels for "Möbius cyclobutadiene" by setting up and solving the usual secular determinant. Check your answer against that obtained using the circle mnemonic trick. Compare the resonance energy with that for cyclobutadiene.

13. We are given the following energies and MO expressions for butadiene.

$$X = -1.618; \qquad \psi_1 = 0.3717\chi_1 + 0.6014\chi_2 + 0.6014\chi_3 + 0.3717\chi_4$$

$$X = -0.618; \qquad \psi_2 = 0.6014\chi_1 + 0.3717\chi_2 - 0.3717\chi_3 - 0.6014\chi_4$$

$$X = +0.618; \qquad \psi_3 = 0.6014\chi_1 - 0.3717\chi_2 - 0.3717\chi_3 + 0.6014\chi_4$$

$$X = +1.618; \qquad \psi_4 = 0.3717\chi_1 - 0.6014\chi_2 + 0.6014\chi_3 - 0.3717\chi_4$$

Use this information to calculate (to the second decimal place) the molecular orbital energy levels of the cyclobutadiene. Compare this with the energies obtained by exact Hückel solution of the cyclobutadiene problem.

14. Using the same butadiene MOs as given in Problem 13, derive the approximate energy levels of methylenecyclopropene. Here the 1,3-partial bond orders are used.

15. Again using the butadiene MOs of Problem 13, derive the approximate energy levels of "Möbius cyclobutadiene" and compare your answer with that obtained in Problem 12.

16. What is the effect of a methoxyl group substituted on (a) a carbonium ion center, (b) a free radical center, (c) a carbanion center? That is, which centers are stabilized or destabilized and by what relative amount? Explain this in MO terms: For simplicity simulate an oxygen p orbital by a carbon p-orbital.

References

1. J. C. Slater, *Phys. Rev.* **36,** 57 (1930).
2. R. S. Mulliken, C. A. Rieke, D. Orloff, and H. Orloff, *J. Chem. Phys.* **17,** 1248 (1949).
3. C. A. Coulson, "Valence." Oxford Univ. Press, London and New York, 1961.
4. E. Hückel, *Z. Phys.* **70,** 204 (1931); **76,** 628 (1932).
5. A. A. Frost and B. Musulin, *J. Chem. Phys.* **21,** 572 (1953).
6. C. A. Coulson, *Proc. Roy. Soc.* **A169,** 413 (1939).
7. E. Heilbronner, *Tetrahedron Lett.* 1923 (1964).
8. H. E. Zimmerman, *J. Amer. Chem. Soc.* **88,** 1564 (1966).
9. R. Breslow, J. Brown, and J. J. Gajewski, *J. Amer. Chem. Soc.* **89,** 4383 (1967).
10. M. J. S. Dewar, *Angew. Chem., Int. Ed.* **10,** 761 (1971); *Tetrahedron Suppl.,* **8,** 75 (1966).
11. H. E. Zimmerman, *Accounts Chem. Res.* **4,** 272 (1971).

Chapter 2

INTRODUCTION TO SOME CONCEPTS OF QUANTUM MECHANICS AND THE THEORETICAL BASIS OF THE LCAO-MO METHOD

The preceding chapter dealt with the simplest methods of the Hückel version of the LCAO-MO approach. Justification of these methods was delayed with the conviction that there is a pedagogical advantage in presenting the utility of the method prior to detailing its theoretical basis. For the inquisitive and physically inclined the present chapter, giving the basis of the LCAO-MO method, has been delayed long enough.

2.1 Fundamental Concepts of Quantum Mechanics

2.1a Eigenvalues, Eigenfunctions, and Operators

We can begin by defining the terms *eigenvalues* and *eigenfunctions* which are basic to quantum mechanics. *Eigenvalues* are allowed values of some observable property, for example energy. In fact, throughout Chapter 1 we determined the *energy eigenvalues*, or molecular orbital energies, for electrons in different molecular systems. The logic of the term eigenvalue (characteristic value) is clear once we remember that each molecule was found to have its own characteristic set of MO energy levels or energy eigenvalues; that is, only very particular energies were allowed. The energy is said to be quantized. However, observables besides energy may be quantized and therefore have sets of allowed values or eigenvalues; presently we are most interested in energy.

An *eigenfunction* is a function describing the system with a given eigenvalue. Our LCAO molecular orbital expressions are eigenfunctions, since

44

each ψ_i describes the state of an electron having the corresponding eigenvalue X_i or MO energy. We see that for each eigenvalue there corresponds an eigenfunction. We are most interested in eigenfunctions describing the allowed states of an electron; such eigenfunctions are termed orbitals. We use the symbol χ_i and term atomic orbital to describe the state of an electron confined to an atom, while the symbol ψ_i is reserved for molecular eigenfunctions (MOs) describing the state of an electron in a molecule.

Finally, before proceeding we need one further definition. An *operator* is a symbol prefixing some variable and signifying that the particular operation is to be performed on the variable. The operator (d/dx) signifies "take the first derivative with respect to x of ...". The operator (d^2/dx^2) similarly specifies "take the second derivative of." It is the variable to the right of the operator which is operated upon.

2.1b   A Basic Postulate of Quantum Mechanics

A fundamental postulate is used to obtain the eigenvalues and corresponding eigenfunctions for some observable property of a given system (e.g., energy and orbital of an electron). This postulate is given by

$$(O_p)\Phi_i = k_i\Phi_i \qquad\qquad 2.1\text{-}1$$

Here (O_p) is an operator, Φ is the function describing the system (as an orbital), and k_i is a constant corresponding to an allowed value of some observable property of the system (as energy). This equation then in effect states that for every observable property of a system described by the function Φ, there can be found an operator (O_p) such that when (O_p) is performed on an allowed value of Φ, Φ_i (i.e., the eigenfunction), one obtains back Φ_i multiplied by a constant k_i, where k_i is the eigenvalue, or allowed value, corresponding to the eigenfunction Φ_i.

In principle, all we have to do is to select the proper operator (O_p) corresponding to the observed property we wish to obtain, and then to solve Eq. 2.1-1 for the allowed values (the k_i's) of the observable; in the process of obtaining these eigenvalues we also obtain the eigenfunctions (the Φ_i's) describing the system in its allowed states. Table 2.1-1 gives some examples of observables and their corresponding quantum mechanical operators. Inspection of Table 2.1-1 reveals that physical properties expressed classically in terms of coordinates x, y, and z have quantum mechanical operators which correspond exactly to the classical expression; such physical properties are position in space (x, y, z, r), relative position in space (r_{12}), and potential energies of electrons. We note, however, that the momentum and kinetic energy operators are different from their classical counterparts. The operator equivalent of momentum $(P = mv)$ is

TABLE 2.1-1

SOME OBSERVABLES AND THEIR QUANTUM MECHANICAL OPERATORS[a,b]

Observable	Classical expression	QM operator
Position in space	x (similarly[b]: y and z)	x (similarly[b]: y and z)
Momentum of a particle of mass m	$P = mv$	$(h/2\pi i)\,(d/dx)$
Kinetic energy of a particle of mass m	$T = \frac{1}{2}mv^2$ $= (1/2m)P^2$	$-(h^2/8m\pi^2)\,(d^2/dx^2)$
Potential energy of an electron near a nucleus of effective positive charge Z	$V = -Ze^2/r$	$-Ze^2/r$
Potential energy of two interacting electrons (1 and 2)	$V = e^2/r_{12}$	e^2/r_{12}
Total energy of an electron	$E = T + V_{\text{tot}}$	$\mathcal{H} \equiv -(h^2/8m\pi^2)\,(d^2/dx^2)$ $+ V_{\text{tot}}$

[a] h is Planck's constant, v is velocity, i is $\sqrt{-1}$, r is the distance of the electron from the nucleus, r_{12} is the distance between two electrons, and e is the charge of an electron.

[b] In expressions involving coordinates where only the x component has been listed, if motion in all directions is considered, then one must add equivalent terms in which y and z replace x.

$(h/2\pi i)\,(d/dx)$. To obtain the kinetic energy operator we replace each P in the classical expression $(1/2m)P^2$ by the momentum operator. This gives $\cdot(1/2m)\,(h/2\pi i)\,(d/dx)\,(h/2\pi i)\,(d/dx)$. Since $(d/dx)\,(d/dx)$ means "differentiate twice with respect to x" and $i = \sqrt{-1}$, the kinetic energy operator becomes $-(h^2/8m\pi^2)\,(d^2/dx^2)$ for motion in the x direction. For motion allowed in all directions (d^2/dx^2) is replaced by the sum $(d^2/dx^2) + (d^2/dy^2) + (d^2/dz^2)$.

We can now rewrite the eigenvalue equation 2.1-1 for the special case employing the electron energy operator $\mathcal{H}$ and giving the energy eigenvalues E_i:

$$\mathcal{H}\Phi_i = E_i\Phi_i \qquad\qquad 2.1\text{-}2$$

Equation 2.1-3 is the equivalent one in which the energy operator is written explicitly (here V_{tot} is the potential energy characteristic of the specific system considered):

$$[-(h^2/8m\pi^2)(d^2/dx^2 + d^2/dy^2 + d^2/dz^2) + V_{\text{tot}}]\Phi_i = E_i\Phi_i \qquad 2.1\text{-}3$$

This is the form of the Schrödinger equation most commonly given in undergraduate textbooks.

A second postulate of quantum mechanics, which can be considered to be a corollary to that already given, states that the value (O_b) of the observable property corresponding to the operator (O_p) when a system is described by the function ψ is given by*

$$(O_b) = \frac{\int \psi(O_p)\psi \, dx \, dy \, dz}{\int \psi^2 \, dx \, dy \, dz} \qquad\qquad 2.1\text{-}4$$

Here (O_b) represents the measured value of the observable. In the special case of interest now, the energy of any orbital is given by

$$E = \frac{\int \psi \mathcal{3C} \psi \, dx \, dy \, dz}{\int \psi^2 \, dx \, dy \, dz} \qquad\qquad 2.1\text{-}5$$

This will correctly give an orbital's energy even when the orbital ψ is not an eigenfunction and additionally even when it is not normalized. The integration in these equations is performed over all space, that is, from

$$x = -\infty \text{ to } +\infty, \quad y = -\infty \text{ to } +\infty \quad \text{and} \quad z = -\infty \text{ to } +\infty$$

If the wavefunction or orbital is normalized, the denominator becomes unity. If, additionally, the orbital is an eigenfunction, Eq. 2.1-5 becomes

$$E_i = \int \psi_i \mathcal{3C} \psi_i \, dx \, dy \, dz \qquad\qquad 2.1\text{-}6$$

where ψ_i is an eigenfunction and E_i is its energy. This equation can be considered to be the integrated form of the Schrödinger equation (Eq. 2.1-2), for if we multiply each side of 2.1-2 by the eigenfunction Φ_i (symbol used in 2.1-2 instead of ψ_i) and then integrate over all space, we obtain

$$\int \Phi_i E_i \Phi_i \, dx \, dy \, dz = E_i \int \Phi_i^2 \, dx \, dy \, dz = E_i = \int \Phi_i \mathcal{3C} \Phi_i \, dx \, dy \, dz \quad 2.1\text{-}6'$$

which is the equivalent of 2.1-6. In this, we have made use of the facts that E_i is a constant and may be removed from under the integral sign and that Φ_i is normalized and as a result the integration of its square affords unity.

2.2 The Variation Method; Minimization of LCAO-MO Energy to Give the Secular Equation and Secular Determinant

We shall now apply the preceding to finding the eigenvalues and eigenfunctions (here allowed energies and orbitals, respectively) for an electron

---

* Throughout this discussion we are assuming that the wavefunctions, or orbitals, used are real (i.e., noncomplex).

placed into the π system of an organic molecule. For mathematical simplicity we begin with a three-atom molecule. We shall assume that the eigenfunction is of the LCAO form, that is,

$$\psi = c_1\chi_1 + c_2\chi_2 + c_3\chi_3 \qquad\qquad 2.2\text{-}1$$

However, the molecular orbital given by Eq. 2.2-1 will be the desired eigenfunction only if c_1, c_2, and c_3 are correctly chosen. We use Eq. 2.1-5 to obtain the energy of the molecular orbital, because we have not assumed it to be normalized and an eigenfunction. The energy of ψ will depend on the choice of c_1, c_2, and c_3, but we desire that choice which will minimize the orbital energy. It will be found that the eigenfunctions correspond to energy minima for bonding MOs.

We begin by substituting the LCAO-MO expression of Eq. 2.2-1 into the energy expression of 2.1-5, cross multiplying in the process to obtain

$$E \int (c_1\chi_1 + c_2\chi_2 + c_3\chi_3)^2 \, dv$$

$$= \int (c_1\chi_1 + c_2\chi_2 + c_3\chi_3)\,\mathcal{H}(c_1\chi_1 + c_2\chi_2 + c_3\chi_3) \, dv \qquad 2.2\text{-}2$$

where dv represents the volume element $dx\,dy\,dz$. We may now expand the squared term on the left and the product on the right; in so doing, we must make certain to keep the $\mathcal{H}$ operator in its present order, that is, operating only on terms to the right of itself. We obtain

$$E\left[c_1{}^2 \int \chi_1{}^2 \, dv + c_2{}^2 \int \chi_2{}^2 \, dv + c_3{}^2 \int \chi_3{}^2 \, dv \right.$$

$$\left. + 2c_1c_2 \int \chi_1\chi_2 \, dv + 2c_1c_3 \int \chi_1\chi_3 \, dv + 2c_2c_3 \int \chi_2\chi_3 \, dv \right]$$

$$= c_1{}^2 \int \chi_1\mathcal{H}\chi_1 \, dv + c_2{}^2 \int \chi_2\mathcal{H}\chi_2 \, dv + c_3{}^2 \int \chi_3\mathcal{H}\chi_3 \, dv$$

$$+ c_1c_2 \int \chi_1\mathcal{H}\chi_2 \, dv + c_1c_2 \int \chi_2\mathcal{H}\chi_1 \, dv + c_1c_3 \int \chi_1\mathcal{H}\chi_3 \, dv$$

$$+ c_1c_3 \int \chi_3\mathcal{H}\chi_1 \, dv + c_2c_3 \int \chi_2\mathcal{H}\chi_3 \, dv + c_2c_3 \int \chi_3\mathcal{H}\chi_2 \, dv \quad 2.2\text{-}3$$

It is convenient to use shorthand notation for integrals of the type which appear in Eq. 2.2-3, since these occur in many usages. The integrals are of four types:

$$H_{rr} = \int \chi_r \mathfrak{K} \chi_r \, dv \qquad \text{is termed the } Coulomb \ integral,$$

$$H_{rs} = \int \chi_r \mathfrak{K} \chi_s \, dv \qquad \text{where } \chi_r \text{ and } \chi_s \text{ are different atomic orbitals, is called the } resonance \ integral,$$

$$S_{rr} = \int \chi_r{}^2 \, dv \qquad \text{is a normalization integral, equal to one if the atomic orbitals used are normalized,}$$

$$S_{rs} = \int \chi_r \chi_s \, dv \qquad \text{is termed the } overlap \ integral.$$

Prior to substituting this symbolism into Eq. 2.2-3, let us consider briefly the nature of the four integrals. The integral H_{rr} can be considered to give the energy of an electron in the (isolated*) atomic orbital. H_{rs}, the resonance integral, is the two-center analog of H_{rr}, and can be considered to give the stabilization resulting from overlap and interaction of the two atomic orbitals χ_r and χ_s with the electron being allowed to distribute itself between both orbitals rather than being confined to one. It is shown later that $H_{rs} = H_{sr}$. S_{rs}, as the designation "overlap integral" implies, is a measure of the extent to which the atomic orbitals χ_r and χ_s overlap in space. To the extent that χ_r and χ_s are simultaneously large in certain regions in space, the integral S_{rs} will also be large. If χ_r is always small

---

* This is only approximately true, since $\mathfrak{K}$ in this derivation is the energy operator for the entire molecular orbital Ψ rather than just the atomic orbital χ_r. Thus

$$\int \chi_r \mathfrak{K} \chi_r \, dv = \int \chi_r [d^2/dx^2 + d^2/dy^2 + d^2/dz^2 - Ze^2/r_r - Ze^2/r_s - Ze^2/r_t] \chi_r \, dv$$

$$= \int \chi_r \mathfrak{K}_r \chi_r \, dv + \int \chi_r [-e^2 Z/r_s - e^2 Z/r_t] \chi_r \, dv = E_r + L.$$

Thus the operator $\mathfrak{K}$ contains in addition to $\mathfrak{K}_r$ (the operator corresponding to the atomic orbital χ_r), the extra terms $-e^2/r_s$ and $-e^2/r_t$ which are the potential energy contributions due to attraction of the electron by nuclei s and t. Thus the integral $H_{rr} = \int \chi_r \mathfrak{K} \chi_r \, dv$ gives the energy E_r of (an electron in) atomic orbital χ_r plus an increment L. The absolute value of L is, however, small, since when χ_r is large (i.e., near atom r), then r_s and r_t (the distances from atoms s and t) are also large and their reciprocals are small. All this is equivalent to saying that the Hamiltonian for the MO is not quite the Hamiltonian operator for a single atom.

where χ_s is large and vice versa, then S_{rs} will be small. S_{rr} is the normalization integral. Since χ_r^2 gives the probability of an electron in χ_r being found (i.e., its electron density) at a particular point in space, the integral S_{rr} gives the total electron density. If χ_r is properly normalized, S_{rr} will thus equal one.

Returning now to our derivation, we rewrite Eq. 2.1-9 using the symbolism for the four types of integrals. In doing this we use the fact that $H_{rs} = H_{sr}$. We obtain

$$E[c_1^2 S_{11} + c_2^2 S_{22} + c_3^2 S_{33} + 2c_1 c_2 S_{12} + 2c_1 c_3 S_{13} + 2c_2 c_3 S_{23}]$$

$$= c_1^2 H_{11} + c_2^2 H_{22} + c_3^2 H_{33} + 2c_1 c_2 H_{12} + 2c_1 c_3 H_{13} + 2c_2 c_3 H_{23} \qquad 2.2\text{-}4$$

Now all of the integrals are fixed quantities characteristic of the geometry of the molecule being considered. The energy E of the system is a function of the variables c_1, c_2, and c_3. To obtain the first secular equation we partially differentiate Eq. 2.2-4 implicitly with respect to c_1, keeping c_2 and c_3 constant. We obtain*

$$(\partial E/\partial c_1)[c_1^2 S_{11} + c_2^2 S_{22} + c_3^2 S_{33} + 2c_1 c_2 S_{12} + 2c_1 c_3 S_{13} + 2c_2 c_3 S_{23}]$$

$$+ E[2c_1 S_{11} + 2c_2 S_{12} + 2c_3 S_{13}] = 2c_1 H_{11} + 2c_2 H_{12} + 2c_3 H_{13} \qquad 2.2\text{-}5$$

For an energy minimum $(\partial E/\partial c_1) = 0$ and the first term drops out. Dividing through by 2 and grouping terms, we find

$$(H_{11} - E S_{11})c_1 + (H_{12} - E S_{12})c_2 + (H_{13} - E S_{13})c_3 = 0 \qquad 2.2\text{-}6$$

which is the first of three secular equations which can be obtained. The second and third are obtained similarly from Eq. 2.2-4 by partially differentiating with respect first to c_2 to give the second secular equation (2.2-7) and then with respect to c_3 to give the third (2.2-8):

$$(H_{21} - E S_{21})c_1 + (H_{22} - E S_{22})c_2 + (H_{23} - E S_{23})c_3 = 0 \qquad 2.2\text{-}7$$

$$(H_{31} - E S_{31})c_1 + (H_{32} - E S_{32})c_2 + (H_{33} - E S_{33})c_3 = 0 \qquad 2.2\text{-}8$$

Now if the three secular equations 2.2-6, 2.2-7, and 2.2-8 are not to have the trivial solution $c_1 = 0$, $c_2 = 0$, $c_3 = 0$, then the following condition must be met†:

---

* In this differentiation we remember that the derivative of a product $d(uv)/dx$ is given by $u(dv/dx) + v(du/dx)$.

† A corollary of Cramer's rule states that a necessary condition for a nontrivial solution of a set of simultaneous, linear equations is the disappearance of the determinant of the coefficients of the unknown variables. In the present instance the unknown variables are c_1, c_2, and c_3 while the coefficients of the variables are the terms in parentheses in the secular equations.

$$
\begin{array}{c}
 \quad \chi_1 \qquad\qquad\quad \chi_2 \qquad\qquad\quad \chi_3 \\
\begin{array}{c} \chi_1 \\[2ex] \chi_2 \\[2ex] \chi_3 \end{array}
\left|
\begin{array}{ccc}
(H_{11} - ES_{11}) & (H_{12} - ES_{12}) & (H_{13} - ES_{13}) \\[2ex]
(H_{21} - ES_{21}) & (H_{22} - ES_{22}) & (H_{23} - ES_{23}) \\[2ex]
(H_{31} - ES_{31}) & (H_{32} - ES_{32}) & (H_{33} - ES_{33})
\end{array}
\right| = 0
\end{array}
\qquad 2.2\text{-}9
$$

This is termed the secular determinantal equation. Each element of the determinant has subscripts indicating which atomic orbitals appear in that element's resonance and overlap integrals; and the columns and rows are labeled with these interacting orbitals. The reader can readily demonstrate to his satisfaction that the preceding derivation can be applied to larger systems, or to the smaller system ethylene for that matter, and will afford secular equations and secular determinants of exactly the same form. However, there will be as many secular equations and rows and columns of the secular determinant as there are orbitals mixed (i.e., taken in linear combination). Thus, in general the secular equations are

$$
\sum_s (H_{rs} - ES_{rs})c_s = 0 \qquad \text{for} \quad r = 1, 2, 3, \ldots, n \qquad 2.2\text{-}10
$$

and the secular determinantal equation is

$$
|(H_{rs} - ES_{rs})| = 0 \qquad\qquad 2.2\text{-}11
$$

Turning now to the specific case of allyl and Eq. 2.2-9, we find that simplification is possible. Thus H_{11}, H_{22}, and H_{33} can all be taken as equal and designated by the symbol α representing the Coulomb integral for a carbon p orbital. H_{12}, H_{21}, H_{23}, and H_{32} are the resonance integrals reflecting the extent of electronic interaction between two adjacent p orbitals, and all of these can be replaced by the general symbol β. H_{13} and H_{31} can be neglected since these integrals involve nonadjacent p orbitals. S_{11}, S_{22}, and S_{33} are normalization integrals equal to one. S_{12} ($= S_{21}$) and S_{23} ($= S_{32}$) are neglected in the Hückel approximation; these overlap integrals are relatively small (about 0.25). As will be noted later, the results which we obtain here with the "neglect of overlap" assumption can readily be corrected to include overlap.

As a result of these substitutions Eq. 2.2-9 becomes

$$
\begin{array}{c}
 \quad \chi_1 \qquad\quad \chi_2 \qquad\quad \chi_3 \\
\begin{array}{c} \chi_1 \\[2ex] \chi_2 \\[2ex] \chi_3 \end{array}
\left|
\begin{array}{ccc}
(\alpha - E) & \beta & 0 \\[2ex]
\beta & (\alpha - E) & \beta \\[2ex]
0 & \beta & (\alpha - E)
\end{array}
\right| = 0
\end{array}
\qquad 2.2\text{-}12
$$

Now determinant algebra allows us to multiply or divide all elements of any row or column by a constant. The effect of such an operation is to multiply or divide the value of the entire determinant by that constant. In the present instance we shall divide each column by β, giving

$$
\begin{array}{c}
 & \chi_1 & \chi_2 & \chi_3 \\
\chi_1 & (\alpha - E)/\beta & 1 & 0 \\
\chi_2 & 1 & (\alpha - E)/\beta & 1 \\
\chi_3 & 0 & 1 & (\alpha - E)/\beta
\end{array} = 0 \qquad 2.2\text{-}13
$$

Now let us define

$$
X = (\alpha - E)/\beta = (E - \alpha)/(-\beta) = (E - \alpha)/(|\beta|) \qquad 2.2\text{-}14a
$$

and we note that Eq. 2.2-13 becomes the secular determinantal equation of the form utilized in Chapter 1:

$$
\begin{array}{c}
 & \chi_1 & \chi_2 & \chi_3 \\
\chi_1 & X & 1 & 0 \\
\chi_2 & 1 & X & 1 \\
\chi_3 & 0 & 1 & X
\end{array} = 0 \qquad 1.2\text{-}3
$$

The X's are found as diagonal elements; the ones are found where columns and rows headed by adjacent and interacting atomic orbitals intersect. Zeros are found where columns and rows of noninteracting atomic orbitals intersect.

The definition of X as given in Eq. 2.1-20a is important. We see that X is defined as the energy E of the system in excess of α and in units of $|\beta|$, the absolute value of *beta*. Thus in view of the significance attached to $\alpha = H_{rr}$, the energy of an electron in an isolated p orbital is taken as our arbitrary zero of energy. Thus

$$
E = \alpha + X|\beta| \qquad 2.2\text{-}14b
$$

We have now justified the method of obtaining molecular orbital energies given in Chapter 1.

It is of some considerable interest to inquire whether the solutions to the secular determinantal equation (e.g., 2.2-9) really do correspond to energy minima. Thus far in our derivation we merely required that the first derivative of the energy with respect to each of the LCAO coefficients be zero; this could correspond to an energy minimum, a maximum, or an inflection point. Let us return to Eq. 2.2-5 and implicitly partially differentiate again with respect to c_1. A positive second derivative $(\partial^2 E/\partial c_1^2)$ is a requirement for an energy minimum with respect to c_1 when the secular equation is satisfied; similarly the second derivatives with respect to the other LCAO coefficients should be positive. Implicit partial differentiation affords

$$(\partial^2 E/\partial c_1^2)[c_1^2 S_{11} + c_2^2 S_{22} + c_3^2 S_{33} + 2c_1c_2S_{12} + 2c_1c_3S_{13} + 2c_2c_3S_{23}]$$

$$+ 4(\partial E/\partial c_1)[c_1 S_{11} + c_2 S_{12} + c_3 S_{13}] + E[2S_{11}] = 2H_{11}$$

or

$$(\partial^2 E/\partial c_1^2) = \frac{2(H_{11} - ES_{11})}{c_1^2 S_{11} + c_2^2 S_{22} + c_3^2 S_{33} + 2c_1c_2S_{12} + 2c_2c_3S_{23} + 2c_1c_3S_{13}} \qquad 2.2\text{-}15$$

$$(\partial^2 E/\partial c_1^2) = 2(H_{11} - ES_{11}) = 2(\alpha - E) \qquad 2.2\text{-}16$$

or* in terms of X:

$$(\partial^2 X/\partial c_1^2) = -2X \qquad 2.2\text{-}17$$

The denominator of the expression in 2.2-15 is unity since this is just the expansion of $\int \psi^2 \, dv = \int (c_1\chi_1 + c_2\chi_2 + c_3\chi_3)^2 \, dv$ with substitution for the symbols used for the overlap and normalization integrals.† Analogous expressions are obtained for the second derivatives with respect to the other LCAO coefficients.

Thus we have arrived at the interesting answer in Eq. 2.2-17 that the second derivatives $(\partial^2 X/\partial c_r^2)$ will be positive for a molecular orbital having a value of X less than zero. A positive second derivative provides a necessary condition that the extremum we have located is a minimum, and we can state that all bonding MOs (i.e., negative X's) correspond to energy minima. Any deviation from the value of X afforded by the secular equations will raise the energy of the MO. Furthermore, Eq. 2.2-17 reveals that antibonding MOs, where X is positive, correspond to energy maxima. For nonbonding MOs where $X = 0$, we have an inflection or saddle point. In real organic systems, the bonding MOs will be populated largely or completely while a much smaller number of nonbonding and antibonding MOs will be filled. As a result the energy of the total system will be at a minimum.

* This is obtained by substitution of the definition of X into the right-hand side of 2.2-16 and the second derivative of this definition into the left-hand side.

† Although we have not required the original ψ, whose energy we were extremizing to be normalized, we are requiring that it be a properly behaved MO once the correct c's are chosen.

2.3 Justification of the Method of Cofactors for Determining LCAO Coefficients

Consider a secular determinantal equation as

$$\begin{vmatrix} a_{11} & a_{12} & a_{13} & \cdots & a_{1n} \\ \\ a_{21} & a_{22} & a_{23} & \cdots & a_{2n} \\ \\ a_{31} & a_{32} & a_{33} & \cdots & a_{3n} \\ \cdot & & & & \cdot \\ \cdot & & & & \cdot \\ \cdot & & & & \cdot \\ a_{n1} & & & & a_{nn} \end{vmatrix} = 0 \qquad\qquad 2.3\text{-}1$$

where each element $a_{rs} = (H_{rs} - ES_{rs})$. This is then just a convenient shorthand abbreviation of the secular determinantal equation of the form given in 2.2-9. The secular equations from which 2.3-1 derives are

$$a_{11}c_1 + a_{12}c_2 + a_{13}c_3 + \cdots + a_{1n}c_n = 0 \qquad\qquad 2.3\text{-}2a$$

$$a_{21}c_1 + a_{22}c_2 + a_{23}c_3 + \cdots + a_{2n}c_n = 0 \qquad\qquad 2.3\text{-}2b$$

$$a_{31}c_1 + a_{32}c_2 + a_{33}c_3 + \cdots + a_{3n}c_n = 0 \qquad\qquad 2.3\text{-}2c$$

$$a_{n1}c_1 + \qquad\quad \cdots \qquad\quad + a_{nn}c_n = 0 \qquad\qquad 2.3\text{-}2n$$

where these represent equations such as 2.2-6 through 2.2-8. Now let us see to what extent a valid solution for the molecular orbital coefficients is given by

$$c_s = kA_{1s} \qquad\qquad 2.3\text{-}3$$

where c_s is the LCAO coefficient for atom s, A_{1s} represents the cofactor of element a_{1s}, and k is an undetermined constant. Equation 2.3-3 is just a restatement of the method of cofactors, presented in Chapter 1, in which the relative values of the LCAO coefficients were given by the cofactors of the elements of row 1 of the secular determinant.

We can show in the following way that Eq. 2.3-3 does afford a proper solution to all of the secular equations. If we substitute the values of the LCAO coefficients as given by 2.3-3 into the left-hand side of any of the secular equations, for example the rth one, we obtain a quantity

$$L = k(a_{r1}A_{11} + a_{r2}A_{12} + a_{r3}A_{13} + \cdots + a_{rn}A_{1n}) \qquad\qquad 2.3\text{-}4$$

Now it can be seen* that the portion of Eq. 2.3-4 in parentheses is just an expansion by cofactors, using row 1, of the determinant

$$
\begin{vmatrix}
a_{r1} & a_{r2} & a_{r3} & \cdots & a_{rn} \\
a_{21} & a_{22} & a_{23} & \cdots & a_{2n} \\
a_{31} & a_{32} & a_{33} & \cdots & a_{3n} \\
\cdot & & & & \cdot \\
\cdot & & & & \cdot \\
\cdot & & & & \cdot \\
a_{n1} & & & \cdots & a_{nn}
\end{vmatrix} = L/k \qquad\qquad 2.3\text{-}5
$$

This determinant is zero independent of the value of r, that is, regardless of which secular equation was selected for testing the validity of our choice of coefficients. If we selected the first secular equation and hence $r = 1$, then this determinant is identical with the secular determinant of Eq. 2.3-1, which is equal to zero. If r has any value other than unity (i.e., we selected one of the other secular equations), then the determinant above will have its first row identical with one of the succeeding rows, and any determinant having two identical rows (or columns) is equal to zero. As a consequence $L = 0$ and we see that all of the secular equations are satisfied by choosing the LCAO coefficients as prescribed by the method of cofactors.

2.3a   Orthogonality of Eigenfunctions

Two wavefunctions or orbitals are said to be *orthogonal* when the integral of the product is zero. It is general that two molecular orbitals of the same molecule, or any two eigenfunctions of the same operator for that matter, will be orthogonal if they correspond to different eigenvalues. A general proof is now given. We begin by pointing out that operators of interest in molecular orbital theory satisfy the relation

$$
\int \Psi_i \mathcal{3C} \Psi_j \, dv = \int \Psi_j \mathcal{3C} \Psi_i \, dv \qquad\qquad 2.3\text{-}6
$$

These are termed Hermitian operators, and this property is discussed in the next section. Now if both Ψ_i and Ψ_j are eigenfunctions of the operator, and here $\mathcal{3C}$ is the energy operator as an example, then the Schrödinger equation tells us that

$$
\mathcal{3C}\Psi_i = E_i \Psi_i \qquad \text{and} \qquad \mathcal{3C}\Psi_j = E_j \Psi_j \qquad\qquad 2.3\text{-}7a,b
$$

---

* It is necessary to note that the determinant in Eq. 2.3-5 has the same cofactors of its first-row elements as the secular determinant (i.e., in 2.3-1).

Substitution of 2.3-7a into the right-hand side of 2.3-6 and 2.3-7b into the left-hand side affords after extraction of the constants E_i and E_j,

$$E_i \int \Psi_i \Psi_j \, dv = E_j \int \Psi_j \Psi_i \, dv \qquad\qquad 2.3\text{-}8$$

But since E_i and E_j are different eigenvalues, the equality 2.3-8 can be true only if

$$\int \Psi_i \Psi_j \, dv = 0 \qquad\qquad 2.3\text{-}9$$

The orthogonality relationship of Eq. 2.3-9 can be extended to the special case of two LCAO-MOs by letting

$$\Psi_i = c_{1i}\chi_1 + c_{2i}\chi_2 + c_{3i}\chi_3 + \cdots + c_{ni}\chi_n$$

and

$$\Psi_j = c_{1j}\chi_1 + c_{2j}\chi_2 + c_{3j}\chi_3 + \cdots + c_{nj}\chi_n.$$

Then on substitution into Eq. 2.3-9 we obtain

$$\sum_{r,s}^{n} c_{ri}c_{sj} \int \chi_r \chi_s \, dv = \sum_{r,s}^{n} c_{ri}c_{sj}S_{rs} = 0 \qquad\qquad 2.3\text{-}10$$

With the more stringent neglect of overlap assumption in which S_{rs} is zero except for $r = s$, when $S_{rr} = 1$, we obtain

$$\sum_{r} c_{ri}c_{rj} = 0 \qquad\qquad 2.3\text{-}11$$

That is, when a summation is taken of the products of corresponding LCAO coefficients for two nondegenerate MOs, the result is zero.

Another consequence of this proof is that

$$\int \Psi_i \mathcal{3C} \Psi_j \, dv = 0 \qquad\qquad 2.3\text{-}12$$

since

$$\int \Psi_i \mathcal{3C} \Psi_j \, dv = \int \Psi_j E_j \Psi_i \, dv = E_j \int \Psi_i \Psi_j \, dv = 0 \qquad\qquad 2.3\text{-}13$$

This is the basis of the off-diagonal elements being zero in a secular determinant expressed in terms of molecular orbitals (i.e., the determinant becomes diagonalized). Each off-diagonal element, $H_{rs} - ES_{rs}$, becomes composed of two vanishing portions. The first is zero since it is now the integral in 2.3-13 and the second is zero due to orthogonality of the two MOs as in Eq. 2.3-9.

2.4   The Hermitian Character of the Molecular Orbital Operators

The relationship assumed in Eq. 2.3-6 derives from the nature of the operators used in quantum mechanics and more specifically here, in MO theory. Thus, the energy operator we commonly deal with has two parts, a potential energy term of the form $-Ze^2/r_{ij}$ and a kinetic energy term involving second derivatives.

If we had only the potential energy component, we would have no doubt about the Hermitian character of the operator, since $-Ze^2/r_{ij}$ is just a multiplier and thus the order of terms under the integral sign is unimportant. That is,

$$\int \Psi_i \frac{Ze^2}{r_{ij}} \Psi_j \, dv = \int \Psi_j \frac{Ze^2}{r_{ij}} \Psi_i \, dv = \int \frac{Ze^2}{r_{ij}} \Psi_i \Psi_j \, dv, \ldots \qquad 2.4\text{-}1$$

However, the Hermitian character of the kinetic energy portion of the operator is not as obvious. For simplicity we will test the operator (d^2/dx^2) since the x, y, and z second derivatives are additive in the operator and if one is Hermitian, the rest will be so also. Thus the question is whether the following equation is valid:

$$\int_{-\infty}^{+\infty} \Psi_i \frac{d^2}{dx^2} \Psi_j \, dx = \int_{-\infty}^{+\infty} \Psi_j \frac{d^2}{dx^2} \Psi_i \, dx \qquad 2.4\text{-}2$$

We note that the integrals have limits allowing x to run from minus infinity to plus infinity, since in principle orbitals may extend this far.

To test 2.4-2 we use the method of integration by parts. We remember that

$$\int_{l_1}^{l_2} u \, dw = [uw]_{l_1}^{l_2} - \int_{l_1}^{l_2} w \, du \qquad 2.4\text{-}3$$

For the left-hand side of Eq. 2.4-2 we allow Ψ_i to be u and $(d^2\Psi_j/dx^2) \, dx$ to be dw. This gives us

$$\int_{-\infty}^{+\infty} \Psi_i \underbrace{\frac{d^2}{dx^2} \Psi_j \, dx}_{} = \left[\Psi_i \frac{d\Psi_j}{dx} \right]_{-\infty}^{+\infty} - \int_{-\infty}^{+\infty} \frac{d\Psi_j}{dx} \frac{d\Psi_i}{dx} \, dx \qquad 2.4\text{-}4\text{a}$$

$$\begin{array}{ccccccc} \uparrow & \uparrow & & \uparrow & \uparrow & & \uparrow & \uparrow \\ u & dw & & u & w & & w & du \end{array}$$

Similar treatment of the right-hand side of 2.4-2, however with Ψ_j being u and $(d^2\Psi_i/dx^2) \, dx$ being dw, affords

$$\int_{-\infty}^{+\infty} \Psi_j \frac{d^2}{dx^2} \Psi_i \, dx = \left[\Psi_j \frac{d\Psi_i}{dx} \right]_{-\infty}^{+\infty} - \int_{-\infty}^{+\infty} \frac{d\Psi_j}{dx} \frac{d\Psi_i}{dx} \, dx \qquad 2.4\text{-}4\text{b}$$

In each of Eqs. 2.4-4a and 2.4-4b, we find that the term in brackets is zero. Thus when evaluating this term at its limits a reasonable wavefunction will vanish at plus or minus infinity (as will the derivative). The second term is seen to be the same for the two equations, since the integrals differ only in order of multiplication of two derivatives. Hence we have proved the equality in Eq. 2.4-2 and this leads to completion of the proof of 2.3-6.

2.5   Rank of Secular Determinants as Affecting Determination of Coefficients

In Chapter 1 it was noted that the method of cofactors could not be used to obtain LCAO-MO coefficients for degenerate MOs. In order to explain the basis of this it is necessary to define rank. The rank of a determinant is the size of the largest subdeterminant one can obtain by striking out rows and columns such that the subdeterminant is nonzero. If we have a secular determinant of order (e.g.) 3 (as in cyclopropenyl), we see immediately that the rank cannot also be 3, since the secular determinant by definition is zero. If by deleting one row and one column, we can obtain a nonzero 2×2 subdeterminant, then the rank will be 2. In the case of the secular determinant for cyclopropenyl, all 2×2 determinants become zero when we substitute in the value of $X = +1$. This was the reason the ratio of cofactors could not be used to give the LCAO-MO coefficients. The problem here is that the rank of the cyclopropenyl secular determinant is 1 and not 2 when degenerate eigenvalues are substituted in. Thus, all cofactors will vanish.

More generally, the rank of secular determinants of order n will be $n - 1$ for nondegenerate eigenvalues and $n - 1 - m$ where there are m degeneracies, i.e., m degenerate pairs. The proof of this is left for the reader to consider in connection with later discussions of diagonalization of secular determinants and matrices.

Suggested Reading

W. Kauzmann, "Quantum Chemistry." Academic Press, New York, 1957.

H. Eyring, J. Walter, and G. Kimball, "Quantum Chemistry." Wiley, New York, 1944.

L. Pauling and E. B. Wilson, "Introduction to Quantum Mechanics." McGraw-Hill, New York, 1935.

F. L. Pilar, "Elementary Quantum Chemistry." McGraw-Hill, New York, 1968.

Chapter 3

THE USE OF MOLECULAR SYMMETRY FOR
SIMPLIFICATION OF SECULAR DETERMINANTS;
INTRODUCTION TO GROUP THEORY

At the end of Chapter 1 it became clear that the difficulty of direct solution of secular determinants increases rapidly with the size of the molecule studied and a mathematical impasse is quickly reached. When the molecular system has symmetry, as is frequently the case, this impasse can be delayed by the use of group theory. However, the approach of this chapter is not to present group theory immediately but rather to develop the use of symmetry properties more gradually and then to demonstrate the equivalence of group theory to the methods employed.

3.1 Conversion of Secular Determinants Expressed in Terms of Atomic Orbitals into Secular Determinants Expressed in Terms of Group (Symmetry) Orbitals

3.1a Method of Addition and Subtraction of Rows and Columns

For a molecule which has an element of symmetry such as a plane, there is a useful method for simplification of the secular determinant. Ethylene is the first case to be considered. As before the secular determinant is written in terms of the atomic orbitals which then label the rows and columns. The simplification involves a first step of writing a new determinant in which each pair of columns of the original secular determinant headed by equivalently located atomic orbitals has been added to give one new column and then subtracted to give a second new column. In the same way, in a second step, the rows are added and subtracted.

Original Secular Determinant

$$
\begin{array}{c}
\quad\; \chi_1 \;\; \chi_2 \\
\begin{array}{c} \chi_1 \\ \chi_2 \end{array}
\left|
\begin{array}{cc}
X & 1 \\
1 & X
\end{array}
\right| = 0
\end{array}
$$

Result of Step 1 (Column Add–Sub)

$$
\begin{array}{c}
\quad (\chi_1 + \chi_2) \quad\; (\chi_1 - \chi_2) \\
\begin{array}{c} \chi_1 \\ \chi_2 \end{array}
\left|
\begin{array}{cc}
X + 1 & X - 1 \\
1 + X & 1 - X
\end{array}
\right|
\end{array}
$$

Result of Step 2 (Row Add–Sub)

$$
\begin{array}{c}
\qquad\qquad (\chi_1 + \chi_2) \quad (\chi_1 - \chi_2) \\
\begin{array}{c} (\chi_1 + \chi_2) \\ (\chi_1 - \chi_2) \end{array}
\left|
\begin{array}{c:c}
2X + 2 & 0 \\
\hdashline
0 & 2X - 2
\end{array}
\right| = 0
\end{array}
$$

Hence

$$
\begin{array}{c}
(\chi_1 + \chi_2) \\
(\chi_1 + \chi_2) |\; 2X + 2 \;| = 0 \qquad or \qquad X = -1
\end{array}
$$

and

$$
\begin{array}{c}
(\chi_1 - \chi_2) \\
(\chi_1 - \chi_2) |\; 2X - 2 \;| = 0 \qquad or \qquad X = +1
\end{array}
$$

Addition–subtraction operations as performed above do not alter the equality of a determinant to zero.* These operations afford a secular determinant which can be dissected into two 1×1 determinants. In general, a determinant which can be separated into such nonoverlapping blocks of smaller determinants is equal to the product of these smaller determinants. Then each determinant equaling zero provides a solution to the secular equation. In the present instance, since a 1×1 determinant of a quantity is equal to the quantity itself, the addition–subtraction method has led directly to the final solutions of $X = -1$ and $X = +1$. Furthermore, the group orbitals labeling the column and row of each of the 1×1 secular determinants are the corresponding unnormalized molecular orbitals.

In the utilization of the addition–subtraction method it is important to

---

* Each addition–subtraction operation (e.g., on two columns) has the effect of multiplying the determinant by 2. However, since the determinant is equal to zero, multiplication by a constant is of no concern. The overall operation is equivalent to a "similarity transformation" and multiplication by a constant; this matrix equivalent is considered subsequently.

group the added columns together and separately from the subtracted columns. The rows should follow the same sequence as the columns. The group orbitals heading the added columns can be termed "symmetric" group orbitals while those labeling the subtracted columns can be designated "antisymmetric." The order in which the symmetric orbitals occur is unimportant as long as the same order is observed for columns and rows. The same is true of the separately grouped antisymmetric orbitals.

A second example is found in the treatment of the allyl species (I).

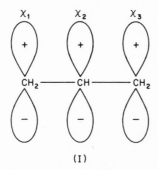

(I)

Here there is a plane of symmetry bisecting the molecule through atom 2 and χ_2. Since χ_1 and χ_3 are the equivalently located atomic orbitals, the columns and rows headed by these will be added and subtracted to give a secular determinant expressed in terms of symmetric and antisymmetric group orbitals. Since χ_2 has no similarly situated orbital, it is not subjected to addition–subtraction operations. Since it is bisected by the plane of symmetry, it is symmetrical with respect to this plane and is grouped with the symmetrical group orbitals. The operations are then

Original Secular Determinant

$$
\begin{array}{c|ccc}
 & \chi_1 & \chi_2 & \chi_3 \\
\hline
\chi_1 & X & 1 & 0 \\
\chi_2 & 1 & X & 1 \\
\chi_3 & 0 & 1 & X
\end{array} = 0
$$

Result of Column Operations

$$
\begin{array}{c|ccc}
 & (\chi_1 + \chi_3) & \chi_2 & (\chi_1 - \chi_3) \\
\hline
\chi_1 & X & 1 & X \\
\chi_2 & 2 & X & 0 \\
\chi_3 & X & 1 & -X
\end{array} = 0
$$

Result of Further Row Operations

$$
\begin{array}{cc}
& \begin{array}{ccc} (\chi_1 + \chi_3) & \chi_2 & (\chi_1 - \chi_3) \end{array} \\
\begin{array}{c} (\chi_1 + \chi_3) \\[20pt] \chi_2 \\[20pt] (\chi_1 - \chi_3) \end{array} &
\left|\begin{array}{ccc}
2X & 2 & 0 \\
2 & X & 0 \\
0 & 0 & 2X
\end{array}\right| = 0
\end{array}
$$

Hence

$$
\begin{array}{cc}
& \begin{array}{cc} (\chi_1 + \chi_3) & \chi_2 \end{array} \\
\begin{array}{c} (\chi_1 + \chi_3) \\[16pt] \chi_2 \end{array} &
\left|\begin{array}{cc}
2X & 2 \\
2 & X
\end{array}\right| = 0 \quad and \quad X = \pm\sqrt{2}
\end{array}
$$

for the symmetric orbitals

$$
\begin{array}{cc}
& \begin{array}{c} (\chi_1 - \chi_3) \end{array} \\
(\chi_1 - \chi_3) &
\left|\begin{array}{c} 2X \end{array}\right| = 0 \quad and \quad X = 0
\end{array}
$$

for the antisymmetric orbital

In the present instance addition–subtraction operations have broken the third-order secular determinant into a 2×2 composed of symmetric orbitals, plus a 1×1 of the antisymmetric orbital $\chi_1 - \chi_3$ whose energy is $X = 0$ and is the nonbonding MO found by the earlier direct approach. Whenever a 1×1 determinant is obtained, not only does this yield the energy of an orbital directly but also the heading of the determinant is more than just a group orbital—it is the unnormalized molecular orbital having the energy given by the determinant. Contrariwise, when simplification by symmetry has left a determinant headed by several orbitals of the same symmetry, as in the case of the second-order determinant above, these orbitals must be mixed further to afford a solution. This is done by solving the determinantal equation for the energy, here $X = \pm\sqrt{2}$ and then using the method of cofactors to determine the coefficients weighting the orbitals heading the columns and rows of the simplified secular determinant. Thus the second-order secular determinant obtained for the allyl species has its elements expressed in terms of the two orbitals $\phi_s = (\chi_1 + \chi_3)$ and χ_2. The molecular orbitals whose energies are given by this determinant will have the form

$$\psi = c_s\phi_s + c_2\chi_2 = c_s(\chi_1 + \chi_3) + c_2\chi_2 \qquad 3.1\text{-}1$$

and the coefficients obtained from the cofactors of the first-row elements of the second-order determinant are then

Coefficient	Cofactor in general	For $X = -\sqrt{2}$	For $X = +\sqrt{2}$
c_s	X	$-\sqrt{2}$	$+\sqrt{2}$
c_2	-2	-2	-2

and the unnormalized MOs are

$$\psi_{-\sqrt{2}} = \sqrt{2}\phi_s + 2\chi_2 = \sqrt{2}\chi_1 + \sqrt{2}\chi_3 + 2\chi_2 \qquad 3.1\text{-}2$$

$$\psi_{+\sqrt{2}} = \sqrt{2}\phi_s - 2\chi_2 = \sqrt{2}\chi_1 + \sqrt{2}\chi_3 - 2\chi_2 \qquad 3.1\text{-}3$$

These are equivalent to the orbitals given in Eqs. 1.3-13 except for a constant factor of $\sqrt{2}$. The normalized MOs are presently obtained by division by the square root of the sum of the unnormalized coefficients squared (i.e., $\sqrt{8}$):

$$\psi_{-\sqrt{2}} = \tfrac{1}{2}\chi_1 + \tfrac{1}{2}\chi_3 + (1/\sqrt{2})\chi_2 \qquad 3.1\text{-}4$$

$$\psi_{+\sqrt{2}} = \tfrac{1}{2}\chi_1 + \tfrac{1}{2}\chi_3 - (1/\sqrt{2})\chi_2 \qquad 3.1\text{-}5$$

3.1b Some Comments on Symmetry and Antisymmetry of Group Orbitals

In the preceding discussion, the subtracted group orbitals have been referred to as antisymmetric while the groups consisting of added atomic orbitals have been termed symmetric. The significance of this designation can be conveyed either pictorially or algebraically. Thus inspection of $(\chi_1 + \chi_3)$ (Fig. 3.1-A) and $(\chi_1 - \chi_3)$ (Fig. 3.1-B) indicates that the orbital $(\chi_1 + \chi_3)$ is symmetrically disposed about the plane of symmetry. However, $(\chi_1 - \chi_3)$ is antisymmetric with respect to the plane since for every point on one side of the plane there is an equal but negative value of the orbital found at the equivalently located point on the other side of the plane. χ_2 is symmetric; for this reason it was grouped with the summed symmetric orbital.

FIG. 3.1-A. The group orbital $(\chi_1 + \chi_3)$.   FIG. 3.1-B. The group orbital $(\chi_1 - \chi_3)$.

For this to be seen algebraically, we need to define a symmetry operator σ, which means "reflect in the plane of symmetry." This symmetry operator will now be applied to the group orbitals of interest:

If

$$\phi_s = \chi_1 + \chi_3 \quad \text{and} \quad \phi_a = \chi_1 - \chi_3 \qquad 3.1\text{-}6$$

then

$$\sigma\phi_s = \sigma(\chi_1 + \chi_3) = \sigma\chi_1 + \sigma\chi_3 = \chi_3 + \chi_1 = \phi_s \qquad 3.1\text{-}7$$

$$\sigma\phi_a = \sigma(\chi_1 - \chi_3) = \sigma\chi_1 - \sigma\chi_3 = \chi_3 - \chi_1 = -\phi_a \qquad 3.1\text{-}8$$

$$\sigma\chi_2 = \chi_2 \qquad 3.1\text{-}9$$

since

$$\sigma\chi_1 = \chi_3 \quad \text{and} \quad \sigma\chi_3 = \chi_1 \qquad 3.1\text{-}10$$

Equation 3.1-10 follows from inspection of Fig. 3.1-A and recognition that performance of the operation σ on χ_1 converts it into χ_3 and the same operation on χ_3 transforms it into χ_1.

One further point is noteworthy. Equations 3.1-7 and 3.1-8 reveal that our group orbitals ϕ_s and ϕ_a are eigenfunctions of the operator σ, for application of this operator to each of these results in the same orbital, ϕ_s and ϕ_a respectively, multiplied by a constant $+1$ in the case of ϕ_s and by a constant -1 in the case of ϕ_a. Thus the net result of the symmetry operator on a group orbital is to multiply the orbital by its eigenvalue $+1$ if the orbital is symmetric and -1 if it is antisymmetric.

3.1c   Application to Cyclobutadiene

Cyclobutadiene is a molecule which has two planes of symmetry (cf. Fig. 3.1-C) and hence two reflection operations σ and σ'. The present use of symmetry in the addition–subtraction method is of interest not only because of the great simplification of the secular determinant but also because

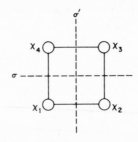

F_{IG}. 3.1-C

as noted on pages 34 and 35 we are unable to obtain the LCAO-MO coefficients from the fourth-order secular determinant but are able to derive these presently.

The fourth-order secular determinant can be first simplified by adding and subtracting rows and columns* headed by atomic orbitals symmetrically placed about the σ plane.

Original Secular Determinant

$$
\begin{array}{c@{\quad}c@{\quad}c@{\quad}c@{\quad}c}
 & \chi_1 & \chi_2 & \chi_3 & \chi_4 \\
\chi_1 & X & 1 & 0 & 1 \\
\chi_2 & 1 & X & 1 & 0 \\
\chi_3 & 0 & 1 & X & 1 \\
\chi_4 & 1 & 0 & 1 & X
\end{array} = 0 \qquad 3.1\text{-}11a
$$

Addition–Subtraction of Columns

$$
\begin{array}{c@{\quad}c@{\quad}c@{\quad}c@{\quad}c}
 & (\chi_1 + \chi_4) & (\chi_2 + \chi_3) & (\chi_1 - \chi_4) & (\chi_2 - \chi_3) \\
\chi_1 & (X+1) & 1 & (X-1) & 1 \\
\chi_2 & 1 & (X+1) & 1 & (X-1) \\
\chi_3 & 1 & (1+X) & -1 & (1-X) \\
\chi_4 & (1+X) & 1 & (1-X) & -1
\end{array} = 0 \quad 3.1\text{-}11b
$$

Second Step: Addition–Subtraction by Rows

$$
\begin{array}{c@{\quad}c@{\quad}c@{\quad}c@{\quad}c}
 & (\chi_1 + \chi_4) & (\chi_2 + \chi_3) & (\chi_1 - \chi_4) & (\chi_2 - \chi_3) \\
(\chi_1 + \chi_4) & (2X+2) & 2 & 0 & 0 \\
(\chi_2 + \chi_3) & 2 & (2X+2) & 0 & 0 \\
(\chi_1 - \chi_4) & 0 & 0 & (2X-2) & 2 \\
(\chi_2 - \chi_3) & 0 & 0 & 2 & (2X-2)
\end{array} = 0 \quad 3.1\text{-}11c
$$

It is noted that there result two second-order secular determinants—one expressed in terms of group orbitals symmetric with respect to the horizontal plane of symmetry and the other antisymmetric with respect to the

* To avoid circumlocution we refer to addition (subtraction) of rows (columns). Addition (subtraction) of the corresponding elements of the rows (columns) is actually meant.

horizontal plane. These 2×2 determinants can now be solved for the molecular orbital energies and the LCAO coefficients. We can begin by dividing all columns by 2. Then

$$
\begin{array}{c}
 \quad (\chi_1 + \chi_4) \quad (\chi_2 + \chi_3) \\
\begin{array}{c}
(\chi_1 + \chi_4) \\
(\chi_2 + \chi_3)
\end{array}
\left|
\begin{array}{cc}
(X + 1) & 1 \\
1 & (X + 1)
\end{array}
\right| = 0
\end{array}
\quad
\begin{array}{l}
\text{or} \quad (X + 1)^2 = 1 \quad\quad \text{3.1-12} \\
\text{or} \quad X = -1 \pm 1 \\
\text{and} \quad X = -2, 0 \\
\text{for the horizontally sym-} \\
\text{metric orbitals}
\end{array}
$$

The weighting of these two horizontally symmetric orbitals in the LCAO molecular orbitals corresponding to $X = -2$ and 0 is obtained by the method of cofactors:

LCAO coefficients	General expression	For $X = -2$	For $X = 0$
c_{14}	$(X + 1)$	-1	$+1$
c_{23}	-1	-1	-1

MO energy	LCAO-MO wavefunction (unnormalized)	Normalized MO expression
$X = -2$	$\psi_1 = -1(\chi_1 + \chi_4) - 1(\chi_2 + \chi_3)$	$\psi_1 = \tfrac{1}{2}(\chi_1 + \chi_2 + \chi_3 + \chi_4)$ 3.1-13
$X = 0$	$\psi_2 = 1(\chi_1 + \chi_4) - 1(\chi_2 + \chi_3)$	$\psi_2 = \tfrac{1}{2}(\chi_1 - \chi_2 - \chi_3 + \chi_4)$ 3.1-14

Similarly, for the horizontally antisymmetric orbitals

$$
\begin{array}{c}
 \quad (\chi_1 - \chi_4) \quad (\chi_2 - \chi_3) \\
\begin{array}{c}
(\chi_1 - \chi_4) \\
(\chi_2 - \chi_3)
\end{array}
\left|
\begin{array}{cc}
(X - 1) & 1 \\
1 & (X - 1)
\end{array}
\right| = 0
\end{array}
\quad
\begin{array}{l}
\text{or} \quad (X - 1)^2 = 1 \\
\text{or} \quad X = 1 \pm 1 \\
\text{and} \quad X = 0, +2 \quad\quad \text{3.1-15}
\end{array}
$$

The method of cofactors is then applied to these MOs to give

LCAO-MO coefficient	General expression	For $X = 0$	For $X = 2$
c_{14}	$(X - 1)$	-1	$+1$
c_{23}	-1	-1	-1

MO energy	LCAO-MO wavefunction (unnormalized)	Normalized MO expression
0	$\psi_3 = -1(\chi_1 - \chi_4) - 1(\chi_2 - \chi_3)$	$\psi_3 = \tfrac{1}{2}(\chi_1 + \chi_2 - \chi_3 - \chi_4)$ 3.1-16
2	$\psi_4 = 1(\chi_1 - \chi_4) - 1(\chi_2 - \chi_3)$	$\psi_4 = \tfrac{1}{2}(\chi_1 - \chi_2 + \chi_3 - \chi_4)$ 3.1-17

However, we have utilized only one of the two planes of symmetry for simplification. We could have employed this further symmetry to decompose each of the second-order determinants into two 1×1 determinants. Thus addition–subtraction by columns and rows of the determinant in 3.1-12 gives 3.1-18 while similar treatment of 3.1-15 affords 3.1-19:

$$
\begin{array}{c|c:c|}
 & \psi_1 & \psi_2 \\
 & (\chi_1 + \chi_2 + \chi_3 + \chi_4) & (\chi_1 - \chi_2 - \chi_3 + \chi_4) \\
\hline
(\chi_1 + \chi_2 + \chi_3 + \chi_4) & (2X + 4) & 0 \\
\hdashline
(\chi_1 - \chi_2 - \chi_3 + \chi_4) & 0 & 2X \\
\end{array} = 0
$$

3.1-18

$$
\begin{array}{c|c:c|}
 & \psi_3 & \psi_4 \\
 & (\chi_1 + \chi_2 - \chi_3 - \chi_4) & (\chi_1 - \chi_2 + \chi_3 - \chi_4) \\
\hline
(\chi_1 + \chi_2 - \chi_3 - \chi_4) & 2X & 0 \\
\hdashline
(\chi_1 - \chi_2 + \chi_3 - \chi_4) & 0 & (2X - 4) \\
\end{array} = 0
$$

3.1-19

Each of the resulting first-order secular determinants affords one of the MO energies found from 3.1-12 and 3.1-15. Additionally, the orbitals heading these determinants no longer occur together with other group orbitals in a secular determinant and thus no further mixing is required. This means that these orbitals are more than just group orbitals; they correspond to the molecular orbitals as determined earlier by the method of cofactors. Inspection of the symmetry properties of the four MOs reveals that ψ_1 is symmetric with respect to both reflection operations, σ and σ'. ψ_2 is symmetric with respect to σ but antisymmetric with respect to σ'. ψ_3 is antisymmetric with respect to σ but symmetric with respect to σ'. ψ_4 is antisymmetric with respect to both operations. These molecular orbitals are pictured in Fig. 3.1-D. The final first-order secular determinants could have been obtained directly by a single process of taking linear combinations of rows and columns. Thus column 1 would be taken as the sum of all four columns of the original fourth-order determinant; column 2 as the sum of columns 1 and 4 minus columns 2 and 3; column 3 as the sum of columns 1 and 2 minus columns 3 and 4; column 4 as the sum of columns 1 and 3 minus columns 2 and 4. The rows would then be added and subtracted in the same way. This process leads to decomposition into the same four first-order secular determinantal equations as derived from the stepwise decomposition.

Two aspects of the preceding deserve further emphasis and generalization. First, in general, group orbitals of different symmetry will not mix

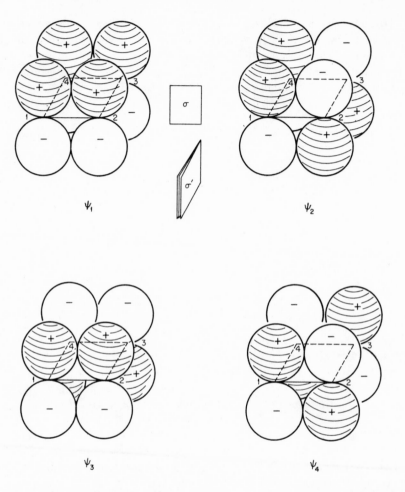

FIG. 3.1-D.   Cyclobutadiene molecular orbitals: ψ_1, symmetric with respect to σ and σ'; ψ_2, symmetric with respect to σ, antisymmetric with respect to σ'; ψ_3, antisymmetric with respect to σ, symmetric with respect to σ'; ψ_4, antisymmetric with respect to σ and σ'.

in a secular determinant. That is, all elements corresponding to the intersection of a row and a column headed by (e.g., group) orbitals of different symmetry will be zero. By reformulating a secular determinant (originally expressed in terms of atomic orbitals) in terms of group orbitals it is thus possible to transform this determinant into one composed of blocks of smaller determinants and equal to the product of these smaller determinants.

Second, different molecular orbitals will not mix in a secular determinant

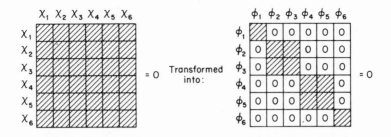

F<small>IG</small>. 3.1-E   Hypothetical sixth-order secular determinant transformed into two second-order and two first-order determinants. The χ's are atomic orbitals and the ϕ's are group orbitals, all of different symmetry type except ϕ_2 and ϕ_3, which are the same, and ϕ_4 and ϕ_5, which are the same.

either.* Thus, if one had a "crystal ball" which told him the LCAO form of the molecular orbitals of a molecule, one could add and subtract columns and rows (or multiples of columns and rows as needed) so as to transform the original secular determinant into one expressed in terms of combinations of the original atomic orbitals corresponding to the LCAO form of the molecular orbitals. Then a determinant having first-order determinants along the diagonal would result, since every off-diagonal element would correspond to the intersection of a column and a row headed by different MOs. Since each first-order determinant resulting directly affords the energy of the corresponding MO, such a "diagonalization process" is the goal of secular determinant simplification. Symmetry provides an only partially effective "crystal ball" leading to only partial diagonalization of the secular determinant (see, for example, Fig. 3.1-E).

3.1d   Direct Formulation of Symmetry Determinants

Thus far we have used the addition–subtraction device for transforming secular determinants expressed in terms of atomic orbitals into secular determinants formulated in terms of group, or symmetry, orbitals. The addition–subtraction device had pedagogical value and is of interest. Nevertheless, there is a simple method of writing down the transformed determinant directly. To obtain any element (a_{rs}) of a secular determinant expressed in terms of group orbitals, one multiplies the group orbital ϕ_r heading the row in which the element of interest appears by the group orbital ϕ_s heading the column.

---

* A difficulty arises when a degenerate pair is considered. However, even here no mixing occurs when these are expressed in proper form. The nature of the "proper form" is discussed subsequently.

(1)   For every squared term (i.e., χ_r^2) in the product $\phi_r\phi_s$ we include one X in that element a_{rs}.

(2)   For every cross product of two adjacent and interacting atomic orbitals, we add a one.

(3)   For every cross product of noninteracting atomic orbitals, we put in a zero.

Elements at the intersection of columns and rows headed by group orbitals of different symmetry can be written as zero without using these rules; however, it may be worthwhile for the reader to convince himself that the rules arrive at the same prediction.

Furthermore, since a secular determinant expressed in terms of group orbitals automatically decomposes itself into smaller determinants, each one of which is expressed in terms of group orbitals of only one symmetry type, one need not deal with the secular determinant as a whole but rather one can deal separately with each of these smaller secular determinants.

The method just given can now be applied to molecules discussed previously from other viewpoints. In the case of ethylene (II) we can write

II

two group orbitals $\phi_s = \chi_1 + \chi_2$ and $\phi_a = \chi_1 - \chi_2$ which are, respectively, symmetric and antisymmetric with respect to the plane perpendicular to sigma bond 1–2. The 2×2 determinant written in terms of these is then

$$
\begin{array}{c c}
 & \begin{array}{cc} (\chi_1 + \chi_2) & (\chi_1 - \chi_2) \end{array} \\
\begin{array}{c} (\chi_1 + \chi_2) \\[1.5em] (\chi_1 + \chi_2) \end{array} &
\left|
\begin{array}{c:c}
2X + 2 & 0 \\ \hdashline
0 & 2X - 2
\end{array}
\right| = 0,
\end{array}
\qquad X = -1, \qquad X = +1
$$

For a_{11} the orbital product is

$$(\chi_1 + \chi_2)(\chi_1 + \chi_2) = \chi_1^2 + \chi_2^2 + 2\chi_1\chi_2.$$

There are two squared atomic orbital terms, hence a_{11} includes $2X$. Two cross-product terms of interacting AOs are present; hence a_{11} includes 2.

For a_{22} the orbital product is

$$(\chi_1 - \chi_2)(\chi_1 - \chi_2) = \chi_1^2 + \chi_2^2 - 2\chi_1\chi_2.$$

Again, there are two squared AO terms, hence a_{22} includes $2X$. Minus two cross product terms of interacting AOs requires a_{22} to include -2.

For a_{12} and a_{21} the orbital product is

$$(\chi_1 - \chi_2)(\chi_1 + \chi_2) = \chi_1^2 - \chi_2^2.$$

Plus one and minus one squared term are present, so for a_{12} and a_{21} no X's are included and, no cross-product terms are found.

Since the symmetric and antisymmetric orbitals did not mix, each could have been dealt with separately in a 1×1 determinant.

For allyl (I) the antisymmetric group orbital $\phi_a = \chi_1 - \chi_3$ will not

$$\overset{1}{CH_2} - \overset{2}{CH} - \overset{3}{CH_2}$$

(I)

mix with the two symmetric orbitals χ_2 and $\phi_s = \chi_1 + \chi_3$, and consequently ϕ_a is already an unnormalized MO. To get its energy we merely need to write down the first-order determinant

$$
\begin{array}{c}
(\chi_1 - \chi_3) \\
(\chi_1 - \chi_3) \left| \quad 2X \quad \right| = 0
\end{array}
\quad \text{giving} \quad X = 0 \qquad 3.1\text{-}20
$$

using the rules for writing down the single element. The symmetric orbitals, χ_2 and ϕ_s, have to be mixed in a secular determinant

$$
\begin{array}{c}
\qquad (\chi_1 + \chi_3) \quad \chi_2 \\
(\chi_1 + \chi_3) \left| \begin{array}{cc} 2X & 2 \\[2mm] 2 & X \end{array} \right| = 0 \quad \text{or} \quad 2X^2 - 4 = 0 \quad \text{and} \quad X = \pm\sqrt{2} \\
\chi_2
\end{array}
$$

3.1-21

The LCAO coefficients of the two orbitals mixed, $\phi_s = \chi_1 + \chi_3$ and χ_2, are obtained from this second-order determinant by the method of cofactors:

LCAO coefficient	General expression	For $X = -\sqrt{2}$	For $X = +\sqrt{2}$
c_{13}	X	$-\sqrt{2}$	$+\sqrt{2}$
c_2	-2	-2	-2

MO energy	LCAO-MO wavefunction (unnormalized)	Normalized MO expression	
$-\sqrt{2}$	$\psi_1 = -\sqrt{2}(\chi_1 + \chi_3) - 2\chi_2$	$\psi_1 = \frac{1}{2}\chi_1 + (1/\sqrt{2})\chi_2 + \frac{1}{2}\chi_3$	3.1-22
$+\sqrt{2}$	$\psi_3 = +\sqrt{2}(\chi_1 + \chi_3) - 2\chi_2$	$\psi_3 = \frac{1}{2}\chi_1 - (1/\sqrt{2})\chi_2 + \frac{1}{2}\chi_3$	3.1-23
0	$\psi_2 = \chi_1 - \chi_3$	$\psi_2 = (1/\sqrt{2})\chi_1 - (1/\sqrt{2})\chi_3$	3.1-24

The case of the cyclopropenyl species (III) is analogous. The antisymmetric orbital $\phi_a = \chi_1 - \chi_3$ is written by inspection and the corresponding

$$\overset{\displaystyle\overset{|}{\underset{2}{CH}}}{\underset{\displaystyle HC-\overset{|}{\underset{|}{-}}CH}{\diagup\diagdown}}$$

$$\overset{1}{HC}-\overset{|}{\underset{|}{}}-\overset{3}{CH}$$

(III)

first-order determinant formulated as

$$(\chi_1 - \chi_3)$$
$$(\chi_1 - \chi_3)\,|\,2X - 2\,| = 0 \qquad \text{and} \qquad X = 1 \qquad \text{3.1-25}$$

We note that this is different from the first-order determinant of the allyl system, for presently in the orbital product $\chi_1^2 + \chi_3^2 - 2\chi_1\chi_3$ the term $-2\chi_1\chi_3$ contributes -2 since χ_1 and χ_3 are adjacent and interacting unlike the situation in the allyl case.

The symmetric cyclopropenyl orbitals are χ_2 and $\phi_s = \chi_1 + \chi_3$. Mixed together in a secular determinant these lead to two further eigenvalues:

$$
\begin{array}{c}
\begin{array}{cc} (\chi_1 + \chi_3) & \chi_2 \end{array} \\
\begin{array}{c} (\chi_1 + \chi_3) \\ \\ \chi_2 \end{array}
\begin{vmatrix} (2X + 2) & 2 \\ \\ 2 & X \end{vmatrix} = 0
\end{array}
\quad
\begin{array}{l}
\text{or} \quad 2X^2 + 2X - 4 = 0 \\
\text{or} \quad (X - 1)(X + 2) = 0; \\
\text{thus} \quad X = 1, -2 \qquad \text{3.1-26}
\end{array}
$$

Using the method of cofactors we obtain after normalization

$$\psi_1 = (1/\sqrt{3})\chi_1 + (1/\sqrt{3})\chi_2 + (1/\sqrt{3})\chi_3$$

$$\psi_2 = (1/\sqrt{6})\chi_1 - (2/\sqrt{6})\chi_2 + (1/\sqrt{6})\chi_3$$

The asymmetrical group orbital ϕ_a needs only to be normalized to afford the third MO $\psi_3 = (1/\sqrt{2})\chi_1 - (1/\sqrt{2})\chi_3$. Of these three MOs, because of difficulties due to degeneracy, we were previously able to obtain the LCAO coefficients only for ψ_1 (cf. pp. 34–35). The present treatment has decomposed the original third-order determinant into two smaller determinants, neither one of which contains a degenerate pair; hence the original difficulty has been circumvented.

In the case of cyclobutadiene the present treatment allows a facile solution. The four group orbitals of different symmetry have been given previously (pp. 66–67); these could be written by inspection of Fig. 3.1-F,

	Symmetry with respect to	
Group orbital	σ	σ'
$\chi_1 + \chi_2 + \chi_3 + \chi_4$	Symmetric	Symmetric
$\chi_1 - \chi_2 - \chi_3 + \chi_4$	Symmetric	Antisymmetric
$\chi_1 + \chi_2 - \chi_3 - \chi_4$	Antisymmetric	Symmetric
$\chi_1 - \chi_2 + \chi_3 - \chi_4$	Antisymmetric	Antisymmetric

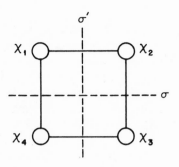

FIG. 3.1-F

using the different combinations of equivalently situated atomic orbitals in such a manner to include all possible combinations of different symmetry. Since these four group orbitals do not mix, they are more than just group orbitals—they are unnormalized MOs. The energy of each is obtained from a first-order determinant. For example,

$$(\chi_1 + \chi_2 + \chi_3 + \chi_4)$$
$$(\chi_1 + \chi_2 + \chi_3 + \chi_4)| \quad 4X + 8 \quad | = 0 \quad \text{and} \quad X = -2$$

The remaining MO energies are similarly obtained as 0, 0, and $+2$. This is clearly less laborious than the addition–subtraction approach of page 65.

The case of benzene is also illustrative of the present approach of formulating the secular determinant directly in terms of group orbitals and then making use of the fact that group orbitals of different symmetry do not mix in a secular determinant. As seen in Fig. 3.1-G benzene has two

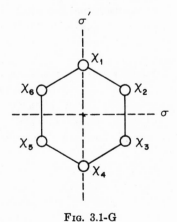

FIG. 3.1-G

perpendicular planes of symmetry, and the corresponding reflection operations are signified by σ and σ'. We could attack the problem by utilizing either one of these symmetry planes alone. Using only σ', we can write down by inspection the group orbitals which are symmetric and antisymmetric with respect to this operation. The symmetric orbitals are χ_1, χ_4, $(\chi_2 + \chi_6)$, and $(\chi_3 + \chi_5)$. The antisymmetric orbitals are $(\chi_2 - \chi_6)$ and $(\chi_3 - \chi_5)$. All six are eigenfunctions of the σ' operator. The four symmetric orbitals have an eigenvalue of $+1$, while the two antisymmetric orbitals have an eigenvalue of -1. The antisymmetric orbitals may be used to write down a second-order secular determinant and the symmetric orbitals lead to a fourth-order secular determinant. Here again we have used the rules given on page 70 to obtain the elements of these determinants:

$$
\begin{array}{c}
\quad\quad\quad (\chi_2 - \chi_6) \quad (\chi_3 - \chi_5) \\
\begin{array}{c} (\chi_2 - \chi_6) \\[1em] (\chi_3 - \chi_5) \end{array}
\left|
\begin{array}{cc}
2X & 2 \\[1em]
2 & 2X
\end{array}
\right| = 0
\end{array}
\qquad \text{3.1-27a}
$$

$$
\begin{array}{c}
\quad\quad\quad \chi_1 \quad \chi_4 \quad (\chi_2 + \chi_6) \quad (\chi_3 + \chi_5) \\
\begin{array}{c} \chi_1 \\[1em] \chi_4 \\[1em] (\chi_2 + \chi_6) \\[1em] (\chi_3 + \chi_5) \end{array}
\left|
\begin{array}{cccc}
X & 0 & 2 & 0 \\[1em]
0 & X & 0 & 2 \\[1em]
2 & 0 & 2X & 2 \\[1em]
0 & 2 & 2 & 2X
\end{array}
\right| = 0
\end{array}
\qquad \text{3.1-27b}
$$

While the second-order determinantal equation 3.1-27a can readily be solved for $X = \pm 1$, the fourth-order equation 3.1-27b is less obviously amenable to facile solution. In actual fact, 3.1-27b can be solved, for example, by expansion by cofactors and thence to a fourth-order polynomial which is found to be readily factored. However, dealing with fourth-order determinants is bothersome.

Had we focused attention on the second symmetry operation σ, we could have written orbitals which are instead symmetric and antisymmetric with respect to this operation. The symmetric orbitals are then $(\chi_1 + \chi_4)$, $(\chi_2 + \chi_3)$, and $(\chi_5 + \chi_6)$. The antisymmetric orbitals are $(\chi_1 - \chi_4)$, $(\chi_2 - \chi_3)$, and $(\chi_5 - \chi_6)$. All of these are eigenfunctions of the σ operator, the former three having eigenvalues of $+1$ and the latter three having eigenvalues of -1. Again we may write down two secular determinants, although still in terms of these symmetry orbitals. This has an advantage since two third-order determinants result and third-order determinants are readily amenable to solution:

$$
\begin{array}{c|ccc}
 & (\chi_1 + \chi_4) & (\chi_2 + \chi_3) & (\chi_6 + \chi_5) \\
\hline
(\chi_1 + \chi_4) & 2X & 2 & 2 \\
(\chi_2 + \chi_3) & 2 & (2X + 2) & 0 \\
(\chi_6 + \chi_5) & 2 & 0 & (2X + 2)
\end{array} = 0 \qquad \text{3.1-28a}
$$

$$
\begin{array}{c|ccc}
 & (\chi_1 - \chi_4) & (\chi_2 - \chi_3) & (\chi_6 - \chi_5) \\
\hline
(\chi_1 - \chi_4) & 2X & 2 & 2 \\
(\chi_2 - \chi_3) & 2 & (2X - 2) & 0 \\
(\chi_6 - \chi_5) & 2 & 0 & (2X - 2)
\end{array} = 0 \qquad \text{3.1-28b}
$$

Expansion of 3.1-28a, after dividing each column by 2, affords $X(X + 1)^2 - 2(X + 1) = (X + 1)(X^2 + X - 2) = (X + 1)(X - 1)(X + 2) = 0$ and $X = -1, +1, -2$. Similar expansion of 3.1-28b gives the MO energies $X = +1, -1, +2$. In the usual fashion the cofactors of row 3* of 3.1-28a could be used to determine the weighting of the group orbitals of this determinant for each of the MO energy levels deriving from this determinant. Similarly, the cofactors of row 1 of the determinant of 3.1-28b give the weighting of the group orbitals of this determinant in the three MOs obtained from the solution of 3.1-28b.

Still easier than either of the two preceding approaches is one that utilizes both planes of symmetry simultaneously. Thus we can write down symmetry orbitals of four types: those symmetric with respect to both σ and σ'; those symmetric with respect to σ but antisymmetric to σ'; those antisymmetric to σ but symmetric to σ'; and finally those antisymmetric with respect to both operations. These are given in Table 3.1-1. An entry of "symmetric" corresponds to an eigenvalue of $+1$ for that group orbital and operator, while "antisymmetric" relates to an eigenvalue of -1. Thus for example, $\sigma(\chi_2 - \chi_3 + \chi_5 - \chi_6) = -1 \cdot (\chi_2 - \chi_3 + \chi_5 - \chi_6)$ as can be demonstrated by carrying out the indicated operation using Fig. 3.1-G to determine the result of each individual σ operation on an atomic orbital; the procedure is the same as demonstrated on pages 63 and 64. Before proceeding to use of these four types of group orbitals, we might pause to question if additional planes of symmetry possessed by the benzene mole-

---

* Row 3 is selected as a second choice when it is observed that the cofactors of row 1 are all zero for some of the eigenvalues. Where a secular determinant contains no degeneracy one can find at least one row whose cofactors afford the coefficients.

TABLE 3.1-1
BENZENE GROUP ORBITALS, SYMMETRY, AND SYMMETRY
EIGENVALUES[a]

	Symmetry with respect to	
Group orbital	σ	σ'
$\chi_1 + \chi_4$	Symmetric (+1)	Symmetric (+1)
$\chi_2 + \chi_3 + \chi_5 + \chi_6$	Symmetric (+1)	Symmetric (+1)
$\chi_2 + \chi_3 - \chi_5 - \chi_6$	Symmetric (+1)	Antisymmetric (−1)
$\chi_1 - \chi_4$	Antisymmetric (−1)	Symmetric (+1)
$\chi_2 - \chi_3 - \chi_5 + \chi_6$	Antisymmetric (−1)	Symmetric (+1)
$\chi_2 - \chi_3 + \chi_5 - \chi_6$	Antisymmetric (−1)	Antisymmetric (−1)

[a] The symmetry eigenvalues are given in parentheses.

cule and not utilized here might be employed to break down our group orbitals into further categories. The answer is that additional planes of reflection are of no use, for the orbitals of Table 3.1-1 can be seen not to be eigenfunctions of such reflection operations; for example, reflection in a plane passing through χ_2 and χ_5 would not convert our group orbitals into themselves or their negatives.

The use of two planes of symmetry allows us now to write down two second-order and two first-order secular determinants. Each of the group orbitals affording a first-order determinant is in fact a final, although unnormalized, MO since it will mix with none of the remaining five group orbitals. Accordingly,

$$ (\chi_2 + \chi_3 - \chi_5 - \chi_6) $$
$$ (\chi_2 + \chi_3 - \chi_5 - \chi_6) | \quad 4X + 4 \quad | = 0 \quad \text{gives } X = -1 \text{ and} $$
$$ \psi_2 = \tfrac{1}{2}(\chi_2 + \chi_3 - \chi_5 - \chi_6) $$

while

$$ (\chi_2 - \chi_3 + \chi_5 - \chi_6) $$
$$ (\chi_2 - \chi_3 + \chi_5 - \chi_6) | \quad 4X - 4 \quad | = 0 \quad \text{gives } X = 1 \text{ and} $$
$$ \psi_5 = \tfrac{1}{2}(\chi_2 - \chi_3 + \chi_5 - \chi_6) $$

as the eigenvalues and normalized MOs deriving from 1×1's. The two second-order determinants, each based on two group orbitals of the same symmetry, lead to the remaining four eigenvalues; for each determinant we use the method of cofactors to obtain the coefficients. For the completely symmetric orbitals:

$$
\begin{array}{c|cc|}
 & (\chi_1 + \chi_4) & (\chi_2 + \chi_3 + \chi_5 + \chi_6) \\
\hline
(\chi_1 + \chi_4) & 2X & 4 \\
(\chi_2 + \chi_3 + \chi_5 + \chi_6) & 4 & 4X + 4 \\
\end{array} = 0
$$

or $X^2 + X - 2 = 0$ and $X = +1, -2$ 3.1-29

Coefficient	Relative general value	For $X = 1$	For $X = -2$
c_{14}	$X + 1$	$+2$	-1
c_{2356}	-1	-1	-1

MO energy	Normalized molecular orbital
-2	$\psi_1 = (1/\sqrt{6})(\chi_1 + \chi_2 + \chi_3 + \chi_4 + \chi_5 + \chi_6)$
$+1$	$\psi_4 = (1/\sqrt{12})(2\chi_1 - \chi_2 - \chi_3 + 2\chi_4 - \chi_5 - \chi_6)$

For the σ symmetric and σ' antisymmetric orbitals:

$$
\begin{array}{c|cc|}
 & (\chi_1 - \chi_4) & (\chi_2 - \chi_3 - \chi_5 + \chi_6) \\
\hline
(\chi_1 - \chi_4) & 2X & 4 \\
(\chi_2 - \chi_3 - \chi_5 + \chi_6) & 4 & 4X - 4 \\
\end{array} = 0
$$

or $X^2 - X - 2 = 0$ and $X = 2, -1$ 3.1-30

Coefficient	Relative general value	For $X = -1$	For $X = 2$
c_{14}	$X - 1$	$+2$	$+1$
c_{2356}	-1	$+1$	-1

MO energy	Normalized molecular orbital
-1	$\psi_3 = (1/\sqrt{12})(2\chi_1 + \chi_2 - \chi_3 - 2\chi_4 - \chi_5 + \chi_6)$
$+2$	$\psi_6 = (1/\sqrt{6})(\chi_1 - \chi_2 + \chi_3 - \chi_4 + \chi_5 - \chi_6)$

We may summarize the results obtained above for the benzene problem in Fig. 3.1-H.

3.2 Matrix Methods for Diagonalizing Secular Determinants and Matrices; The Heisenberg Formulation of Quantum Mechanics

For each secular determinant one can devise a determinant such that the secular determinant will have its columns added and subtracted as

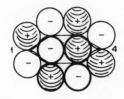

$\chi = 2$

$\psi_6 = (1/\sqrt{6})(\chi_1 - \chi_2 + \chi_3 - \chi_4 + \chi_5 - \chi_6)$

Six nodes

$\chi = 1$

$\psi_5 = 1/2\,(\chi_2 - \chi_3 + \chi_5 - \chi_6)$

Four nodes

$\chi = 1$

$\psi_4 = (1/\sqrt{12})(2\chi_1 - \chi_2 - \chi_3 - 2\chi_4 - \chi_5 - \chi_6)$

$\chi = -1$

$\psi_2 = 1/2\,(\chi_2 + \chi_3 - \chi_5 - \chi_6)$

Two nodes

$\chi = -1$

$\psi_3 = (1/\sqrt{12})(2\chi_1 + \chi_2 - \chi_3 - 2\chi_4 - \chi_5 + \chi_6)$

$\chi = -2$

$\psi_1 = (1/\sqrt{6})(\chi_1 + \chi_2 + \chi_3 + \chi_4 + \chi_5 + \chi_6)$

No nodes (excluding molecular plane)

FIG. 3.1-H.   The six benzene molecular orbitals.

required by symmetry if the secular determinant is postmultiplied by the new determinant. The following provide some examples:

Secular determinant	*Postmultiplying determinant*	*Resulting determinant*

Ethylene

$\chi_1 \quad \chi_2$

$$\begin{vmatrix} X & 1 \\ 1 & X \end{vmatrix} \qquad \begin{vmatrix} 1 & 1 \\ 1 & -1 \end{vmatrix} \qquad \begin{matrix} \chi_1 + \chi_2 & \chi_1 - \chi_2 \end{matrix} \\ \begin{vmatrix} (X+1) & (X-1) \\ (X+1) & (1-X) \end{vmatrix}$$

Allyl

$\chi_1 \quad \chi_2 \quad \chi_3$

$$\begin{vmatrix} X & 1 & 0 \\ 1 & X & 1 \\ 0 & 1 & X \end{vmatrix} \qquad \begin{vmatrix} 1 & 0 & 1 \\ 0 & 1 & 0 \\ 1 & 0 & -1 \end{vmatrix} \qquad \begin{matrix} (\chi_1 + \chi_3) & \chi_2 & (\chi_1 - \chi_3) \end{matrix} \\ \begin{vmatrix} X & 1 & X \\ 2 & X & 0 \\ X & 1 & -X \end{vmatrix}$$

Cyclopropenyl

$$\begin{vmatrix} X & 1 & 1 \\ 1 & X & 1 \\ 1 & 1 & X \end{vmatrix} \qquad \begin{vmatrix} 1 & 0 & 1 \\ 0 & 1 & 0 \\ 1 & 0 & -1 \end{vmatrix} \qquad \begin{vmatrix} (X+1) & 1 & (X-1) \\ 2 & X & 0 \\ (1+X) & 1 & (1-X) \end{vmatrix}$$

Cyclobutadiene

$$\begin{vmatrix} X & 1 & 0 & 1 \\ 1 & X & 1 & 0 \\ 0 & 1 & X & 1 \\ 1 & 0 & 1 & X \end{vmatrix} \quad \begin{vmatrix} 1 & 0 & 0 & 1 \\ 0 & 1 & 1 & 0 \\ 0 & 1 & -1 & 0 \\ 1 & 0 & 0 & -1 \end{vmatrix} \quad \begin{vmatrix} (X+1) & 1 & 1 & (X-1) \\ 1 & (X+1) & (X-1) & 1 \\ 1 & (1+X) & (1-X) & -1 \\ (1+X) & 1 & -1 & (1-X) \end{vmatrix}$$

We note that the postmultiplying determinant is so designed that each column will effect the desired addition and subtraction of columns to give a secular determinant expressed in terms of symmetry orbitals heading the columns. This is effectively the same operation as just adding and subtracting columns as previously. Each column of the postmultiplying determinant has a $+1$ in an element corresponding to an orbital which is needed in the group orbital and needed positively. Minus ones are put in locations in the column where the orbital is needed in minus linear combination. Where an atomic orbital is not needed in the particular group orbital, a zero is entered into that location in the column of the postmultiplying determinant. Thus far, each column of the postmultiplying determinant has had only two entries; this corresponds to use of only one element of symmetry. One could use two perpendicular planes of symmetry, for example, and this would require further entries into the columns.

In similar fashion one could design a premultiplier determinant which added and subtracted rows in the same way. This premultiplying determinant is seen to be the "transpose" of the postmultiplying one; that is, it is obtained by transposing (i.e., exchanging) rows and columns of the postmultiplying determinant. Thus pre- and postmultiplication is seen to simplify the determinant in the same way as addition–subtraction.

We have already noted that if ever an addition–subtraction operation leads to a 1×1 determinant, then this determinant affords a final eigenvalue and the heading of the column and row is a final MO. Thus, if one knew in advance what the LCAO-MO coefficients were for a given problem, one could construct the postmultiplying determinant by using one column for each set of LCAO-MO coefficients corresponding to a given MO. The row vectors of the premultiplying determinant are chosen in the same way. This pre- and postmultiplication then totally diagonalizes the original secular determinant. Each resulting diagonal element has the form $X - A_k$. Since each element equals zero, the A_k's are the eigenvalues.

If the column and row vectors are not normalized, the diagonal elements will have the form $NX - NA_k$, but this does not affect the value of MO energy obtained by setting each 1×1 determinant equal to zero.

One example is the cyclopropenyl problem where

$$
\begin{vmatrix} \dfrac{1}{\sqrt{3}} & \dfrac{1}{\sqrt{3}} & \dfrac{1}{\sqrt{3}} \\[2mm] 0 & \dfrac{1}{\sqrt{2}} & \dfrac{-1}{\sqrt{2}} \\[2mm] \dfrac{2}{\sqrt{6}} & \dfrac{-1}{\sqrt{6}} & \dfrac{-1}{\sqrt{6}} \end{vmatrix}
\cdot
\begin{vmatrix} X & 1 & 1 \\ 1 & X & 1 \\ 1 & 1 & X \end{vmatrix}
\cdot
\begin{vmatrix} \dfrac{1}{\sqrt{3}} & 0 & \dfrac{2}{\sqrt{6}} \\[2mm] \dfrac{1}{\sqrt{3}} & \dfrac{1}{\sqrt{2}} & \dfrac{-1}{\sqrt{6}} \\[2mm] \dfrac{1}{\sqrt{3}} & \dfrac{-1}{\sqrt{2}} & \dfrac{-1}{\sqrt{6}} \end{vmatrix}
$$

$$
= \begin{vmatrix} (X+2) & 0 & 0 \\ 0 & (X-1) & 0 \\ 0 & 0 & (X-1) \end{vmatrix} = 0 \qquad 3.2\text{-}1
$$

We note that in using X's and ones in the Hückel method, we have the negative of all elements of a secular determinant of the form

$$\begin{vmatrix} H_{11} - E & H_{12} & H_{13} \\ H_{21} & H_{22} - E & H_{23} \\ H_{31} & H_{32} & H_{33} - E \end{vmatrix} = 0 \qquad\qquad 3.2\text{-}2$$

but this just changes the sign of the entire determinant and does not change its equality to zero. Here the diagonal H's are zero due to our choice of our energy zero, and our off-diagonal H's are $-1, 0,$ or $+1$ due to our choice of energy units. If one wanted, however, to use the exact form of 3.2-2 in terms of X's and ones, we would need to put $-X$'s on the diagonal and -1 for every off-diagonal element corresponding to (plus–plus) or (minus–minus) interaction. Since solution in this way only requires extra effort we will retain the use of $+X$'s when secular determinants are solved in the Hückel approximation. However, the pre- and postmultiplication method will work equally well on a secular determinant of the form 3.2-2.

It is now of considerable help to use a different terminology in discussing the same problem. For this it is necessary to define a "matrix." A matrix is an array of numbers and superficially looks like a determinant. Unlike a determinant it need not be a square array. A matrix does not have a definite value in itself but rather merely provides a means of storing information. Each element of a matrix carries some significance depending on the user's problem. Thus one could use a matrix to store calendar dates, LCAO-MO coefficients, or other data. Presently we will be interested in an **H** matrix whose elements H_{rs} are the resonance integrals. Additionally, a **C** matrix is needed; this will have its columns storing the LCAO-MO coefficients. Finally, we need an **E** matrix in which the diagonal elements are the eigenvalues and there are zeros everywhere else.

In dealing with matrices, we note that these follow the same rules for multiplication which are used in multiplying determinants. One example is the following:

$$\tilde{\mathbf{C}}[\mathbf{H} - \mathbf{E}]\mathbf{C} = 0 \qquad\qquad 3.2\text{-}3$$

where $\tilde{\mathbf{C}}$ is the transpose of **C**.

The array $[\mathbf{H} - \mathbf{E}]$ is the same as that in the secular determinant of Eq. 3.2-2, and Eq. 3.2-3 merely states that if we pre- and postmultiply the matrix $[\mathbf{H} - \mathbf{E}]$ by the transposed and original untransposed arrays of coefficients, we will get a matrix which consists of all zeros. This is identical to what we have been doing with determinants except that previously we have been leaving the E's as symbols and then solving the 1×1 determinants left along the diagonal. Expansion of Eq. 3.2-3 gives

$$\tilde{C}HC - \tilde{C}EC = 0$$

But it can be readily shown that the order of multiplication of a diagonal matrix as E does not affect the final value and thus $\tilde{C}EC = E\tilde{C}C$. Also, the coefficient matrices C presently used are said to be orthonormal; that is, $\tilde{C}C = 1$ where 1 is the unit matrix with ones along the diagonal and zeros elsewhere. This is equivalent to saying that for orthonormal matrices the inverse C^{-1} is given by the transpose $\tilde{C}$. Thus

$$\tilde{C}HC = E \qquad\qquad 3.2\text{-}4$$

This pre- and postmultiplication by an orthonormal matrix and its transpose (or inverse) thus diagonalizes the H matrix and is called a similarity transformation.* Equation 3.2-4 gives the Heisenberg formulation of quantum mechanics and is seen to be exactly equivalent to the secular equation solution of the Schrödinger equation.

In order to use Eq. 3.2-4 exactly one needs to have normalized MOs to construct C and its transpose. As noted above we can fill in the elements of the H matrix as minus ones or zeros since these are the resonance integrals. Thus each nonzero element $H_{rs} = \beta$ or -1 units of $|\beta| = -\beta$.

Some applications are

Allyl

$$
\begin{array}{cccc}
\tilde{C} & H & C & E
\end{array}
$$

$$
\begin{bmatrix}
\dfrac{1}{2} & \dfrac{1}{\sqrt{2}} & \dfrac{1}{2} \\[2mm]
\dfrac{1}{\sqrt{2}} & 0 & \dfrac{-1}{\sqrt{2}} \\[2mm]
\dfrac{1}{2} & \dfrac{-1}{\sqrt{2}} & \dfrac{1}{2}
\end{bmatrix}
\begin{bmatrix}
0 & -1 & 0 \\
-1 & 0 & -1 \\
0 & -1 & 0
\end{bmatrix}
\begin{bmatrix}
\dfrac{1}{2} & \dfrac{1}{\sqrt{2}} & \dfrac{1}{2} \\[2mm]
\dfrac{1}{\sqrt{2}} & 0 & \dfrac{-1}{\sqrt{2}} \\[2mm]
\dfrac{1}{2} & \dfrac{-1}{\sqrt{2}} & \dfrac{1}{2}
\end{bmatrix}
=
\begin{bmatrix}
-\sqrt{2} & 0 & 0 \\
0 & 0 & 0 \\
0 & 0 & +\sqrt{2}
\end{bmatrix}
$$

---

* Such a similarity transformation does not change the sum of the diagonal elements of the matrix (i.e., the "trace" of the matrix). Therein lies the proof of the statement made earlier that the rank of a nondegenerate secular determinant is $n - 1$ and that the rank is diminished by 1 for each degeneracy. Thus, we can include $a - E$ (or $a - X$) in each diagonal element to use the usual formulation. In a determinant having no degeneracy, only one element of the final, diagonalized determinant is zero (i.e., $H_{rr} - X = 0$ for $X = H_{rr}$). If two diagonal elements have the same eigenvalue H_{rr}, then two elements will be zero after diagonalization, and so on. Thus, the rank after diagonalization will be $n - 1 -$ (the number of degeneracies). And we know that the rank is not changed by the diagonalization process.

Cyclopropenyl

$$\begin{bmatrix} \dfrac{1}{\sqrt{3}} & \dfrac{1}{\sqrt{3}} & \dfrac{1}{\sqrt{3}} \\[2ex] \dfrac{1}{\sqrt{2}} & 0 & \dfrac{-1}{\sqrt{2}} \\[2ex] \dfrac{2}{\sqrt{6}} & \dfrac{-1}{\sqrt{6}} & \dfrac{-1}{\sqrt{6}} \end{bmatrix} \begin{bmatrix} 0 & -1 & -1 \\ -1 & 0 & -1 \\ -1 & -1 & 0 \end{bmatrix} \begin{bmatrix} \dfrac{1}{\sqrt{3}} & \dfrac{1}{\sqrt{2}} & \dfrac{2}{\sqrt{6}} \\[2ex] \dfrac{1}{\sqrt{3}} & 0 & \dfrac{-1}{\sqrt{6}} \\[2ex] \dfrac{1}{\sqrt{3}} & \dfrac{-1}{\sqrt{2}} & \dfrac{-1}{\sqrt{6}} \end{bmatrix} = \begin{bmatrix} -2 & 0 & 0 \\ 0 & +1 & 0 \\ 0 & 0 & +1 \end{bmatrix}$$

Here again the columns of the **C** matrix are composed of the LCAO-MO coefficients. Taken alone the array of a single MO's coefficients is called a row or column vector. Now instead of using the entire **C** matrix, we could use merely the vector corresponding to one MO. That is,

$$\tilde{\mathbf{c}}_k \mathbf{H} \mathbf{c}_k = E_k \qquad\qquad 3.2\text{-}5$$

Here $\tilde{\mathbf{c}}$ and **c** are the row and column vectors for MO k and E_k is the energy eigenvalue for MO k. Equation 3.2-5 can thus be used conveniently to get the MO energies for a single MO. Some examples are found in the following:

Ethylene; MO 1

$$\tilde{\mathbf{c}}_1 \mathbf{H} \mathbf{c}_1 = \quad [1/\sqrt{2} \ \ 1/\sqrt{2}] \begin{bmatrix} 0 & -1 \\ -1 & 0 \end{bmatrix} \begin{bmatrix} 1/\sqrt{2} \\ 1/\sqrt{2} \end{bmatrix} \quad = -1$$

Allyl; MO 3

$$\tilde{\mathbf{c}}_3 \mathbf{H} \mathbf{c}_3 = \quad [\tfrac{1}{2} \ -1/\sqrt{2} \ \tfrac{1}{2}] \begin{bmatrix} 0 & -1 & 0 \\ -1 & 0 & -1 \\ 0 & -1 & 0 \end{bmatrix} \begin{bmatrix} \tfrac{1}{2} \\ -1/\sqrt{2} \\ \tfrac{1}{2} \end{bmatrix} \quad = +\sqrt{2}$$

Benzene; MO 2

$$\tilde{\mathbf{c}}_2 \mathbf{H} \mathbf{c}_2 = [0 \ \tfrac{1}{2} \ \tfrac{1}{2} \ 0 \ -\tfrac{1}{2} \ -\tfrac{1}{2}] \begin{bmatrix} 0 & -1 & 0 & 0 & 0 & -1 \\ -1 & 0 & -1 & 0 & 0 & 0 \\ 0 & -1 & 0 & -1 & 0 & 0 \\ 0 & 0 & -1 & 0 & -1 & 0 \\ 0 & 0 & 0 & -1 & 0 & -1 \\ -1 & 0 & 0 & 0 & -1 & 0 \end{bmatrix} \begin{bmatrix} 0 \\ \tfrac{1}{2} \\ \tfrac{1}{2} \\ 0 \\ -\tfrac{1}{2} \\ -\tfrac{1}{2} \end{bmatrix} = -1$$

Finally, in connection with the matrix formulation of the eigenvalue problem, it is interesting to note that expansion of the triple vector–matrix–vector multiplication (i.e., $\tilde{\mathbf{c}}\mathbf{H}\mathbf{c}$) for some specific MO gives a series of terms of the form $c_r H_{rs} c_s$, where r and s refer to the basis set of orbitals. The expansion is

$$\tilde{\mathbf{c}}\mathbf{H}\mathbf{c} = c_1{}^2 H_{11} + c_2{}^2 H_{22} + c_3{}^2 H_{33} + \cdots + c_n{}^2 H_{nn} + 2c_1 c_2 H_{12}$$

$$+ \ 2c_1 c_3 H_{13} + \cdots + 2c_2 c_3 H_{23} + \cdots \qquad\qquad 3.2\text{-}6$$

This is more briefly written as

$$\tilde{\mathbf{c}}\mathbf{H}\mathbf{c} = \sum_{r,s} c_r c_s H_{rs} \qquad\qquad 3.2\text{-}7$$

which assumes that $H_{rs} = H_{sr}$, which will be true for the operators used.

We note that the energy of this one MO then consists of (1) one-center contributions which are just Coulomb integrals weighted by the square of the coefficients, that is, by the electron density contribution, and (2) bond order contributions of twice each product of LCAO coefficients multiplied by the resonance integral between the two orbitals. This is in agreement with what we derived earlier by expanding the integrated form of the Schrödinger equation and is really quite equivalent.

3.3 Matrix Methods for Perturbation Calculations

The preceding matrix treatment requires that one know what the LCAO-MO coefficients are in order to solve the matrix equation for MO energies. However, it is possible to use the LCAO-MO coefficients for a molecule closely related to the one under study. Here the **H** matrix for the molecule whose energy is desired should be used. Thus, to get the MO energies for cyclopropenyl, we could use the **C** matrix built from the LCAO-MO coefficients of allyl together with the **H** matrix for cyclopropenyl. This **C** matrix will not diagonalize the **H** matrix but the diagonal elements of the resulting matrix, nevertheless, will be close to those for cyclopropenyl.

It is simpler for purposes of discussion to deal with the vector–matrix–vector treatment in which only one MO is treated at a time. It is seen that we can use the **c** vector for a given molecule together with the **H** matrix for a derived molecule in which there is some new overlap introduced. A reasonable approximation to the MO energy of the derived molecule is obtained. Even simpler is to break the **H** matrix into two parts, $\mathbf{H}_0$ and $\mathbf{H}'$. $\mathbf{H}_0$ is the matrix for the original molecule before introducing the new overlap and $\mathbf{H}'$ is a matrix containing only the elements deriving from the new overlap. That is, $\mathbf{H} = \mathbf{H}_0 + \mathbf{H}'$. But if this **H** matrix is to be pre- and postmultiplied

by the vector corresponding to a molecule before introduction of new bonding, then the vector used corresponds exactly to the molecule from which
H_0 is derived and thus $\tilde{c}_0 H_0 c_0 = E_0$ where E_0 is the exact MO energy for
the original molecule. This can be written as

$$\tilde{c}_0 H c_0 = \tilde{c}_0 (H_0 + H') c_0 = \tilde{c}_0 H_0 c_0 + \tilde{c}_0 H' c_0 = E_0 + E' \qquad 3.3\text{-}1$$

where $E' = \tilde{c}_0 H' c_0$. This means that the energy change on introducing a
geometric and overlap perturbation, i.e., E', can be obtained by pre- and
postmultiplication of the H' matrix by the original c vectors (i.e., c_0).

An example is the case of MO 1 of allyl where we wish to know the energy
change on introducing 1,3-top-top overlap to the point where this is
equivalent in magnitude to normal ethylenic overlap. Written out, this is

$$E' = \begin{bmatrix} \tfrac{1}{2} & 1/\sqrt{2} & \tfrac{1}{2} \end{bmatrix} \begin{bmatrix} 0 & 0 & -1 \\ 0 & 0 & 0 \\ -1 & 0 & 0 \end{bmatrix} \begin{bmatrix} \tfrac{1}{2} \\ 1/\sqrt{2} \\ \tfrac{1}{2} \end{bmatrix} = -0.5$$

Since the original energy of MO 1 of allyl is -1.414 and the energy change
on introducing 1,3-top-top overlap is -0.5, the approximate new energy
is seen to be -1.914. Actually, with an exact Hückel calculation, we would
get -2.000. Thus the approximation we are using is reasonably good. This
method is really equivalent to the bond order approach described earlier
and it is seen that the triplet matrix multiplication does afford twice the
negative of the bond order between the two interacting atoms. The method
is most useful in determining what types of interactions will be energetically
favorable and what overlap will instead raise the energy.

3.4   The Jacobi Method for Diagonalization of Matrices

We have already considered the problem of diagonalizing, or partially
diagonalizing, an H matrix by use of appropriate similarity transformations,
that is, postmultiplying the H matrix by an appropriately designed matrix
Q and premultiplying by the inverse of Q. For complete diagonalization,
we noted that the use of the coefficient matrix C and its transpose was
effective. When final eigenfunctions were still unknown, we found we could
resort to symmetry to design a Q and its inverse.

The present method is one which does not rely on knowledge of either
final eigenfunctions or symmetry but rather proceeds systematically to
eliminate off-diagonal elements. For this approach to be practical for any

H matrix larger than a 2×2, a reiterative computer program is used, and the method is one of the more reliable methods of diagonalizing matrices by computer.

Let us assume an **H** matrix but with just three rows and columns shown explicitly as in Eq. 3.4-1. Our plan is to find an orthonormal matrix **Q** so that post- and premultiplication by **Q** and $\tilde{\mathbf{Q}}$ converts elements h_{rs} and h_{sr} to zero.

$$
\begin{array}{ccc}
\tilde{\mathbf{Q}} & \mathbf{H} & \mathbf{Q}
\end{array}
$$

$$
\begin{bmatrix}
\cos\theta & \sin\theta & 0 \\
-\sin\theta & \cos\theta & 0 \\
0 & 0 & 1
\end{bmatrix}
\begin{bmatrix}
h_{rr} & h_{rs} & h_{ri} \\
h_{sr} & h_{ss} & h_{si} \\
h_{ir} & h_{is} & h_{ii}
\end{bmatrix}
\begin{bmatrix}
\cos\theta & -\sin\theta & 0 \\
\sin\theta & \cos\theta & 0 \\
0 & 0 & 1
\end{bmatrix}
$$

$$
=
\begin{bmatrix}
\begin{pmatrix} h_{rr}\cos^2\theta \\ +(h_{rs}+h_{sr})\sin\theta\cos\theta \\ +h_{ss}\sin^2\theta \end{pmatrix} & \begin{pmatrix} +h_{rs}\cos^2\theta \\ -(h_{rr}-h_{ss})\sin\theta\cos\theta \\ -h_{sr}\sin^2\theta \end{pmatrix} & \begin{pmatrix} h_{ri}\cos\theta \\ +h_{si}\sin\theta \end{pmatrix} \\[4ex]
\begin{pmatrix} -h_{rs}\sin^2\theta \\ -(h_{rr}-h_{ss})\sin\theta\cos\theta \\ +h_{sr}\cos^2\theta \end{pmatrix} & \begin{pmatrix} h_{rr}\sin^2\theta \\ -(h_{rs}+h_{sr})\sin\theta\cos\theta \\ +h_{ss}\cos^2\theta \end{pmatrix} & \begin{pmatrix} -h_{ri}\sin\theta \\ +h_{si}\cos\theta \end{pmatrix} \\[4ex]
\begin{pmatrix} h_{ir}\cos\theta \\ +h_{is}\sin\theta \end{pmatrix} & \begin{pmatrix} -h_{ir}\sin\theta \\ +h_{is}\cos\theta \end{pmatrix} & h_{ii}
\end{bmatrix}
$$

$$3.4\text{-}1$$

Inspection of the **Q** and $\tilde{\mathbf{Q}}$ matrices reveals that they are orthonormal. That is, for any column or row the sums of squares of the elements add up to unity; and, any two columns, when taken as vectors, give zero products when multiplied. This means that $\tilde{\mathbf{Q}}$ is the inverse of **Q**, since the inverse of orthonormal matrices is the transpose. Such pre- and postmultiplication by a matrix and its inverse has been termed earlier a "similarity transformation." Such a similarity transformation of a matrix does not change the eigenvalues of the matrix transformed. Previously, we used the **C** matrix and its transpose to diagonalize the **H** matrix totally. However, presently we will be satisfied with eliminating one element h_{rs} and its symmetrically disposed element h_{sr}.

We wish to find what value of θ will lead to h_{rs} becoming zero. The **H** matrix is symmetrical (i.e., $h_{rs} = h_{sr}$) and the similarity transformation leaves it symmetrical. Thus, annihilating h_{rs} also removes h_{sr}. Setting the second element in the first row of the transformed matrix (i.e., h_{rs} after similarity transformation) equal to zero, affords Eq. 3.4-2; this has assumed $h_{rs} = h_{sr}$:

$$h_{rs}(\cos^2\theta - \sin^2\theta) - (h_{rr} - h_{ss})\sin\theta\cos\theta = 0 \qquad 3.4\text{-}2$$

From this we obtain

$$\frac{\sin(2\theta)}{\cos(2\theta)} = \frac{2h_{rs}}{h_{rr} - h_{ss}} \qquad 3.4\text{-}3a$$

and

$$\cos 2\theta = \frac{\pm(h_{rr} - h_{ss})}{[(h_{rr} - h_{ss})^2 + 4h_{rs}^2]^{1/2}} \qquad 3.4\text{-}3b$$

where the plus sign is chosen in Eq. 3.4-3b for a positive h_{rs} but the minus sign is used if h_{rs} is negative. We can now obtain $\cos\theta$ and $\sin\theta$ from the standard relationships:

$$\cos\theta = \left(\frac{1 + \cos 2\theta}{2}\right)^{1/2} \quad \text{and} \quad \sin\theta = \left(\frac{1 - \cos 2\theta}{2}\right)^{1/2} \qquad 3.4\text{-}4$$

The quantities $\cos\theta$ and $\sin\theta$ are obtained from the value of $\cos 2\theta$ which, in turn, is obtained by use of Eq. 3.4-3b and the elements of the original **H** matrix. These two quantities are then used for the quantities in the transformed matrix as given in Eq. 3.4-1. The two quantities are also employed to evaluate the matrix **Q** used in the similarity transformation.

This first similarity transformation has succeeded only in converting the original h_{rs} and h_{sr} elements to zero. A second similarity transformation is now designed to convert another large off-diagonal element of the **H** matrix to zero. As these similarity transformations are continued, it is observed that elements which were once annihilated, may become nonzero again while zeroing some other off-diagonal element. Nevertheless, the sum of the squares of the off-diagonal elements does systematically diminish and we repeat the operation until all such elements have vanished and only the eigenvalues are left on the diagonal.

While Eq. 3.4-1 uses only a 3×3 matrix for illustration, larger systems are handled in the same way. In each case the transforming matrix **Q** has the cos–sin square matrix so positioned that it occupies rows and columns r and s to eliminate element h_{rs} and the rest of the **Q** matrix has ones along the diagonal. Finally, all of the **Q** matrices used are multiplied in order to obtain a product which must be identical with the desired **C** matrix, since it is this product which has succeeded in complete diagonalization. That is,

$$\mathbf{Q}_1\mathbf{Q}_2\mathbf{Q}_3\cdots\mathbf{Q}_n = \mathbf{C} \qquad\qquad 3.4\text{-}5$$

3.5   More Formal Use of Symmetry by Means of Group Theory

3.5a   *Symmetry Operators*

Hitherto, we have made use of the symmetry operator σ, which means "reflect in a plane of symmetry." Sometimes the specific plane of reflection is indicated when more than one plane of symmetry exists. Thus σ_v signifies "reflect in a vertical plane" while σ_h indicates "reflect in a horizontal plane." There are additional symmetry operators which may be applied to orbitals of interest; these orbitals may be atomic, group, or molecular orbitals. Table 3.5-1 lists a number of these symmetry operators of interest. We shall be most interested in eigenfunctions of such symmetry operators and in building up eigenfunctions of these operators. Thus we note that group orbitals of interest and the final molecular orbitals are eigenfunctions of symmetry operators appropriate to the molecular system under consideration.

TABLE 3.5-1
COMMON SYMMETRY OPERATIONS

Symbol	Operation signified
σ_v	Reflect in a vertical plane of symmetry (going through a vertical axis of symmetry)
σ_h	Reflect in a horizontal plane of symmetry (perpendicular to the vertical axis)
E	Do nothing
C_2	Rotate by $(360°)/2 = 180°$ about a molecular vertical axis of symmetry
C_3	Rotate by $(360°)/3 = 120°$ about a molecular vertical axis of symmetry
C_n	Rotate by $(360°)/n$
i	Invert the molecule using a center of symmetry
S_2	Rotate by $180°$ and then reflect in a horizontal plane (i.e., $\sigma_h C_2$)
R	A general symbol signifying some symmetry operation to be specified

3.5b   Character Tables of Nondegenerate Symmetry Groups

For nondegenerate orbitals, whether they be symmetry or molecular orbitals, we can define the *character* of the orbital under a given symmetry operation as the eigenvalue of the orbital corresponding to the given symmetry operator. For nondegenerate orbitals the eigenvalue and hence the character for a given operation will be either $+1$ or -1, corresponding to the orbital being symmetric or antisymmetric with respect to that operation. The case of degenerate orbitals is considered later.

Each character table consists of a group of symmetry operations and a listing of the characters for all of the conceivable types of symmetry orbitals which might exist. Two common examples are the C_2 group and the C_{2v} group for which the character tables are given in Tables 3.5-2 to 3.5-4. In designation of the symmetry types possibly occurring, A is used for an orbital type which is symmetrical (eigenvalue and character of 1) with

TABLE 3.5-2

The C_2 group representation	Group operations	
	E	C_2
A	1	1
B	1	-1

TABLE 3.5-3

The C_{2v} group representation	Group operations			
	E	C_2	σ_v	σ_v'
A_1	1	1	1	1
A_2	1	1	-1	-1
B_1	1	-1	1	-1
B_2	1	-1	-1	1

TABLE 3-5.4

The C_s group representation	Group operations	
	E	σ_h
A'	1	1
A''	1	-1

respect to the principal rotational operation C_n (e.g., C_2 in the cases of the two group tables given as examples in Tables 3.5-2 and 3.5-3). B is used to signify an orbital type antisymmetric (eigenvalue and character of -1) with respect to the principal rotation. Subscript 1 connotes symmetry (character 1) for the σ_v operation while subscript 2 implies antisymmetry (character -1) with respect to this operator. It will be further noted that the different symmetry types are termed *representations*. A representation consists of the entire sequence of characters for each of the operations and of the given symmetry type.

3.5c Utilization of Character Tables and Symmetry Operators in Formulating Symmetry Orbitals

In any given molecular problem we are interested only in those symmetry operators which are capable of transforming individual atomic orbitals into equivalently situated atomic orbitals in the molecule. We wish to employ just enough of these symmetry operators so that there is an operator available to transform *each* original atomic orbital into *all* of its equivalent atomic orbitals. This is the best criterion in selecting a group character table for use; this selection will minimize the complexity of the manipulations but will ensure full use of molecular symmetry.

Let us use "rectangular cyclobutadiene"* as an example for illustration which is interesting per se (Fig. 3.5-A). Inspection of the C_2 group table and its operators shows that there are insufficient group operators (i.e., only E and C_2) to convert any one AO into all of its equivalents. Thus we can convert χ_1 to χ_3 by the operator C_2 which rotates the molecule by 180°; but we find no operator capable of converting χ_1 into χ_2 or χ_4. We need a more powerful group. The C_{2v} group does satisfy the requirement. χ_1 is

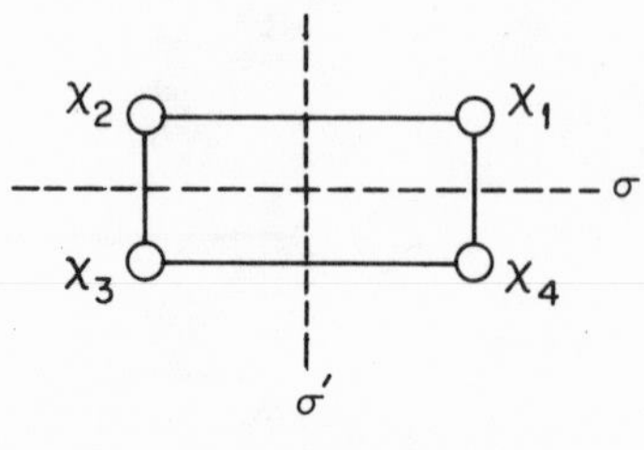

FIG. 3.5-A. The center dot represents the C_2 axis of rotation perpendicular to the molecular plane.

---

* This would be a cyclobutadiene species in which there were alternating bond lengths, and in resonance terminology would correspond to a more heavy weighting of one of the two resonance contributors for cyclobutadiene. Whether in fact square or rectangular cyclobutadiene corresponds to the lower energy situation is a matter of considerable interest.

TABLE 3.5-5

RESULTS OF THE GROUP OPERATIONS ON THE AOS OF
RECTANGULAR CYCLOBUTADIENE

Atomic orbital χ_r	$E\chi_r$	$C_2\chi_r$	$\sigma\chi_r$	$\sigma'\chi_r$
χ_1	χ_1	χ_3	χ_4	χ_2
χ_2	χ_2	χ_4	χ_3	χ_1
χ_3	χ_3	χ_1	χ_2	χ_4
χ_4	χ_4	χ_2	χ_1	χ_3
X_t:	4	0	0	0

converted into itself by the operator E, into χ_3 by the operator C_2, into χ_2 by σ' and into χ_4 by σ. Having selected the C_{2v} group and its operators, we proceed by listing vertically all of the AOs in the molecule. Then in adjacent columns we list the result of each of the group operators on these AOs. For rectangular cyclobutadiene this is carried out in Table 3.5-5. Below each column, at the bottom of the table, is listed the number of AOs unchanged, by the operator heading the column; this will be used later. The following rule (Rule II) will now give all of the needed group orbitals of any specified symmetry type. To obtain a group orbital one uses a linear combination of the atomic orbitals (e.g., given as a row in Table 3.5-5) obtained by the action of all the group operations on any one atomic orbital of the molecule. The coefficients used in this linear combination are just the corresponding characters for the desired symmetry type as selected from the group character table. Thus to obtain an orbital of A_1 symmetry we would use the characters 1, 1, 1, 1 as coefficients; applied to the atomic orbitals of row 1 of Table 3.5-5 this gives the A_1 symmetry orbital $\chi_1 + \chi_3 + \chi_4 + \chi_2$. Applied to the last three rows of the table we obtain the same symmetry orbital written in different order and thus obtain nothing new.

To generate an A_2 group orbital we use the characters for the A_2 representation as taken from the C_{2v} table, namely 1, 1, -1, -1, and use these as coefficients of the atomic orbitals given in row 1 of Table 3.5-5. This gives us the A_2 orbital $\chi_1 + \chi_3 - \chi_4 - \chi_2$. Similarly, use of the characters of the B_1 representation (i.e., 1, -1, 1, -1) as coefficients leads to the B_1 symmetry orbital $\chi_1 - \chi_3 + \chi_4 - \chi_2$; and the characters of the B_2 representation (1, -1, -1, 1) afford $\chi_1 + \chi_2 - \chi_3 - \chi_4$, the B_2 symmetry orbital. As it happens in the present problem, application of the characters of any of the representations to more than the first row of Table 3.5-5 gives us no new group orbitals. Since we have obtained only one orbital of each symmetry type, these will not mix in a secular determinant and are final, although unnormalized, molecular orbitals for the rectangular cyclo-butadiene molecule.*

We can restate the rule just given for obtaining symmetry orbitals by group theory in the following formula. This gives the symmetry orbital ϕ_i of symmetry type i as

Rule II $$\phi_i = \sum_R X_{iR} \cdot R\chi_r$$ 3.5-1

Here X_{iR} is the character of representation (symmetry type) i for opera-tion R. If χ_r is any arbitrarily selected atomic orbital in the molecule, $R\chi_r$ is therefore the atomic orbital resulting from the symmetry operator R acting on χ_r. In the treatment of rectangular cyclobutadiene above we obtained the X_{iR}'s from a row (of the desired symmetry type) of Table 3.5-3 and we obtained the $R\chi_r$'s from (e.g.) row 1 of Table 3.5-5. These products were taken for all symmetry operations of the character table used.

If the reader reflects on this use of group theory to construct the sym-metry orbitals, he will see that it is equivalent in every way to the approach used earlier. Thus Table 3.5-5 is a formal way to obtain a set of atomic orbitals equivalently located in the molecule; and use of the characters of a given representation (symmetry type) as coefficients of these equivalently located AOs merely ensures the desired symmetry or antisymmetry with respect to the operators of the group. As an example, in construction of the A_2 orbital, one of the four terms used is $X_{A_2,\sigma_v} \cdot \sigma_v \chi_1$. The portion $\sigma_v \chi_1$ is equal to χ_4 (cf. Table 3.5-5); this is just the AO located in the molecule in a position equivalent to χ_1. The portion X_{A_2,σ_v} is equal to -1 (cf. Table 3.5-3) and gives χ_4 a negative sign in the summation and ensures antisym-metry with respect to σ_v.

There is another rule (Rule I) of use in constructing symmetry orbitals by use of group theory. This makes use of the total character X_{tR} of the molecular set of atomic orbitals. This total character for each operator R is merely the number of atomic orbitals in the molecule unchanged by that operator. Looking at Table 3.5-5 we see that for the E operator four AOs are unchanged and $X_{tE} = 4$. For each of the remaining operators no AOs are left unaltered, and X_{tC_2}, $X_{t\sigma}$ and $X_{t\sigma'}$ are each zero. These four charac-ters 4, 0, 0, 0 constitute a representation which is said to be reducible.†

---

* We note with interest that these are the same as for the cyclobutadiene molecule itself (cf. p. 72). However, the energy levels will differ in the rectangular and square molecules. Discussion of this problem is postponed despite its intrinsic interest.

† In group theoretical terms the X_{tR}'s are called characters of a reducible representa-tion and the series of these for the different group operators R is called a reducible representation. The set of characters for a given symmetry type (e.g., B_1, etc.) is said to constitute an irreducible representation. Thus a reducible representation may be re-duced, or dissected, into its component irreducible representations.

The C_{2v} group   Group operations

Representation	E	C_2	σ_v	σ_v'
A_1	1	1	1	1
A_2	1	1	−1	−1
B_1	1	−1	1	−1
B_2	1	−1	−1	1
Rectangular cyclobutadiene reducible representation	4	0	0	0

Fig. 3.5-B

These characters, when individually multiplied by the corresponding characters of one of the symmetry types and then the sum divided by the number of operations (the *order* of the group), will give the number of symmetry orbitals of the desired symmetry type. This procedure is most readily remembered by adding a row corresponding to this reducible representation below the C_{2v} group table presently being used (see Fig. 3.5-B). Then these characters are multiplied by the corresponding characters of the symmetry representation of interest; for the A_1 representation we would multiply the pairs of characters connected by arrows in Fig. 3.5-B. The sum of products is then divided here by 4, the order of the C_{2v} group.

Number of A_1 symmetry orbitals in rectangular cyclobutadiene

$$a_{A_1} = \frac{1 \cdot 4 + 1 \cdot 0 + 1 \cdot 0 + 1 \cdot 0}{4} = 1$$

Number of A_2 orbitals

$$a_{A_2} = \frac{1 \cdot 4 + 1 \cdot 0 - 1 \cdot 0 - 1 \cdot 0}{4} = 1$$

Number of B_1 orbitals

$$a_{B_1} = \frac{1 \cdot 4 - 1 \cdot 0 + 1 \cdot 0 - 1 \cdot 0}{4} = 1$$

Number of B_2 orbitals

$$a_{B_2} = \frac{1 \cdot 4 - 1 \cdot 0 - 1 \cdot 0 + 1 \cdot 0}{4} = 1$$

This rule may be formulated algebraically as

Rule I $$a_i \doteq (1/h) \sum_{R}^{h} X_{iR} \cdot X_{tR} \qquad\qquad 3.5\text{-}2$$

where h is the order of the group and a_i is the number of group orbitals of symmetry i. X_{iR} and X_{tR} are the characters of the irreducible and reducible representations, respectively, for operation R.

We see that Rule I does predict that there will be one group orbital of each symmetry type (A_1, A_2, B_1, B_2) in agreement with the finding by Rule II. Normally one would begin by determining the number of orbitals of each type using Rule I and follow this by finding precisely what these symmetry orbitals are.

One further point may be made with regard to the number and type of irreducible representations in the reducible representation provided by the characters of a molecule of interest. The sum of the characters of all of the component irreducible representations will add up to the total character of the reducible representation.* That is,

$$X_{tR} = \sum_i a_i X_{iR} \qquad\qquad 3.5\text{-}3$$

Using this, which will be true for each of the group operations, one can often tell by inspection what combination of symmetry types (irreducible representations) is required to afford the distribution of characters in the reducible representation. Looking at Fig. 3.5-B we can see that the only combination of the A_1, A_2, B_1, and B_2 rows which will give the total characters below the group table is one of each symmetry type. Then the total of E characters adds to 4, the total of C_2 characters gives 0, the total of σ characters gives 0, and the total of σ' characters gives 0. Any other combination would not fit this criterion. For example, if there were two A_1 symmetry orbitals and one each of type B_1 and B_2, the reducible representation would consist of characters 4, 0, 2, 2, which are not the X_t's we obtained in Table 3.5-5.

Another way of considering the dissection of a reducible representation into its irreducible components is to view the representations as vectors. Thus the A_1 representation is the vector $[1\ 1\ 1\ 1]$, while the A_2 representation is the vector $[1\ 1\ -1\ -1]$, and so forth. It is seen that in general the product of any two vectors of different representations (i.e., of differing symmetry) is zero and the product of any vector multiplied by itself is the order of the group, h, here 4 for the C_{2v} group. Also, a reducible representation is a vector, too, and the vector for cyclobutadiene is $\mathbf{\Gamma}_{red} = [4\ 0\ 0\ 0]$. However, this reducible vector is the sum of the four vectors A_1, A_2, B_1, and B_2, each taken once. Accordingly, if one multiplies this reducible vector $\mathbf{\Gamma}_{red}$ by one of the four irreducible vectors, for example the B_1 vector $\mathbf{\Gamma}_{B1}$, he will get contributions to the product only from the B_1 components. Each B_1 component when multiplied by the vector $\mathbf{\Gamma}_{B1}$ then gives a contribution of 4 (i.e., the order of the group C_{2v}). If this multiplication gives

---

* If any representation occurs more than once, it is included this number of times.

4, there will be one B_1 irreducible vector in the reducible sum Γ_{red}. If the product were to turn out to be 8, this would mean that there were two B_1 irreducible representations included in the reducible one. A product of 12 would mean three B_1 components, and so on.

This can be stated in the form of what is a general proof for Rule I of group theory. First, the reducible vector in a general case can be said to be the sum of irreducible component vectors:

$$\Gamma_{red} = a_1\Gamma_1 + a_2\Gamma_2 + a_3\Gamma_3 + \cdots + a_i\Gamma_i \qquad 3.5\text{-}4$$

where each a_j is the number of vectors of the jth symmetry type in the summation. If we now multiply this sum of vectors by one of the irreducible vectors, say Γ_i, only one of the irreducible vectors on the right-hand side of Eq. 3.5-4 will give a nonzero product, and this is $a_i h$. Thus we get

$$\Gamma_{red} \cdot \Gamma_i = 0 + 0 + 0 \cdots + a_i h \qquad 3.5\text{-}5$$

since $\Gamma_i \cdot \Gamma_i = h$. Therefore, we can solve for the number of vectors of the ith variety a_i to obtain

$$a_i = \Gamma_{red} \cdot \Gamma_i (1/h) \qquad 3.5\text{-}6$$

But this can be seen to be just a vector language restatement of Rule I (note Eq. 3.5-2).

We shall now consider a second molecule amenable to treatment by the C_{2v} group, namely 1,4-dehydrobenzene (IV), in resonance language (Fig. 3.5-C). Any less complex group than C_{2v} will not convert each of the atomic

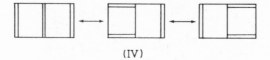

(IV)

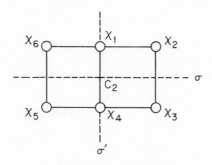

Fig. 3.5-C

TABLE 3.5-6

Atomic orbital	Result of group operator			
	E	C_2	σ	σ'
χ_1	χ_1	χ_4	χ_4	χ_1
χ_2	χ_2	χ_5	χ_3	χ_6
χ_3	χ_3	χ_6	χ_2	χ_5
χ_4	χ_4	χ_1	χ_1	χ_4
χ_5	χ_5	χ_2	χ_6	χ_3
χ_6	χ_6	χ_3	χ_5	χ_2
Total character of reducible representation	6	0	0	2

orbitals into all of its equivalent orbitals, while those more complex involve more than the minimum necessary operations.*

Rule I is employed first in order to determine the number of symmetry orbitals of each type. A table detailing the results of the C_{2v} group's operators on all of the AOs is assembled (Table 3.5-6). The characters of the reducible representation are seen to be 6, 0, 0, 2. Equation 3.5-2 now gives the total number of group orbitals (reducible representations) of each symmetry type. Following this first for the A_1 representation whose characters are 1, 1, 1, 1, we obtain the number of A_1 orbitals as

$$a_{A_1} = \tfrac{1}{4}(1\cdot 6 + 1\cdot 0 + 1\cdot 0 + 1\cdot 2) = 2 \qquad 3.5\text{-}7$$

Analogously

$$a_{A_2} = \tfrac{1}{4}(1\cdot 6 + 1\cdot 0 - 1\cdot 0 - 1\cdot 2) = 1 \qquad 3.5\text{-}8$$

$$a_{B_1} = \tfrac{1}{4}(1\cdot 6 - 1\cdot 0 + 1\cdot 0 - 1\cdot 2) = 1 \qquad 3.5\text{-}9$$

$$a_{B_2} = \tfrac{1}{4}(1\cdot 6 - 1\cdot 0 - 1\cdot 0 + 1\cdot 2) = 2 \qquad 3.5\text{-}10$$

Alternatively, we might have come to this conclusion by inspection by noting that it would require this combination of irreducible representations to afford the reducible representation $(6, 0, 0, 2)$ provided by the dehydrobenzene molecule. This is seen in Fig. 3.5-D.

Knowing the number of each type of symmetry orbital, we proceed to

---

* To be precise one would assign the molecule to the C_{2h} group which also includes a horizontal reflection operation using the plane of the molecule. However, the only *additional* interconversions accomplished by this group is to transform upper lobes of p orbitals into lower lobes. This transformation has no value in providing further symmetry orbitals but instead entails considerable extra work.

	Operators of the C_{2v} group					
Representation	E	C_2	σ	σ'		
A_1	1	1	1	1	Two needed	to give totals of characters
A_2	1	1	-1	-1	One needed	equaling those for reducible
B_1	1	-1	1	-1	One needed	representation
B_2	1	-1	-1	1	Two needed	
Reducible representation of molecule equal to total of above	6	0	0	2		

FIG. 3.5-D

determining what these group orbitals are; Rule II is used. We employ the sequence of AOs given in the rows of Table 3.5-6 and apply the characters of the appropriate symmetry representation as coefficients to these AOs. Accordingly, for the one A_2 orbital we use the characters 1, 1, -1, -1 as coefficients. Applied to the first row of Table 3.5-6 this gives $\chi_1 + \chi_4 - \chi_4 - \chi_1$ or zero; thus row 1 of Table 3.5-6 is of no use. Application of the same four characters to row 2 of the table gives $\chi_2 + \chi_5 - \chi_3 - \chi_6$ (better written $\chi_2 - \chi_3 + \chi_5 - \chi_6$), which is the desired single A_2 symmetry orbital. Further application of the same characters to the remaining rows of Table 3.5-6 gives either no orbital (zero) or the same A_2 orbital; hence we see that the prediction of one A_2 orbital by Rule I is fulfilled.

The energy of the A_2 orbital is obtained in the usual fashion as

$$(\chi_2 - \chi_3 + \chi_5 - \chi_6)| \quad \begin{matrix} (\chi_2 - \chi_3 + \chi_5 - \chi_6) \\ 4X - 4 \end{matrix} \quad | = 0 \quad \text{or} \quad X = 1 \qquad 3.5\text{-}11$$

and the normalized MO is

$$\psi_5 = \tfrac{1}{2}(\chi_2 - \chi_3 + \chi_5 - \chi_6) \qquad 3.5\text{-}12$$

Use of the B_1 characters as coefficients for the AOs of Table 3.5-6 gives $\chi_2 - \chi_5 + \chi_3 - \chi_6$ (or better, $\chi_2 + \chi_3 - \chi_5 - \chi_6$) whose energy is given by

$$(\chi_2 + \chi_3 - \chi_5 - \chi_6)| \quad \begin{matrix} (\chi_2 + \chi_3 - \chi_5 - \chi_6) \\ 4X + 4 \end{matrix} \quad | = 0 \quad \text{or} \quad X = -1 \quad 3.5\text{-}13$$

Here the normalized MO is

$$\psi_2 = \tfrac{1}{2}(\chi_2 + \chi_3 - \chi_5 - \chi_6) \qquad 3.5\text{-}14$$

The A_1 and B_2 orbitals give more difficulty, since there are two group orbitals of each symmetry type which must be mixed in second-order determinants. Beginning with the A_1 case, we obtain the first of the two group orbitals by application of the A_1 representation characters 1, 1, 1, 1 to the first row of Table 3.5-6. This gives $\chi_1 + \chi_4 + \chi_4 + \chi_1$, or just $\chi_1 + \chi_4$.* The second A_1 orbital results from application of the characters to row 2 of Table 3.5-6 to yield (after rearrangement of terms) $\chi_2 + \chi_3 + \chi_5 + \chi_6$. The reader can demonstrate that application of the characters to the last four rows of the same table merely gives repetition of the same orbital. The two A_1 orbitals are now mixed in a second-order secular determinant to give

$$
\begin{array}{cc}
 & (\chi_2 + \chi_3 + \chi_5 + \chi_6) \quad (\chi_1 + \chi_4) \\
\begin{array}{c}(\chi_2 + \chi_3 + \chi_5 + \chi_6)\\[20pt](\chi_1 + \chi_4)\end{array}
\left|\begin{array}{cc} (4X+4) & 4 \\[20pt] 4 & (2X+2) \end{array}\right| = 0 & \quad 3.5\text{-}15
\end{array}
$$

This can be expanded in the usual way to give the two energy levels and then the two sets of coefficients weighting the symmetry orbitals. However, when a determinant having a set of rows and columns of the general form

$$
\left|\begin{array}{cc} 2A & B \\ B & A \end{array}\right| = 0 \qquad\qquad 3.5\text{-}16
$$

is encountered, there is a convenient trick. Row 1 and column 1 are individually divided by $\sqrt{2}$. Application of this gives a symmetrical determinant of the form

$$
\left|\begin{array}{cc} A & (1/\sqrt{2})B \\ (1/\sqrt{2})B & A \end{array}\right| = 0 \qquad\qquad 3.5\text{-}17
$$

which can be diagonalized by adding and subtracting rows and columns. With this approach 3.5-15 becomes†

$$
\begin{array}{cc}
 & (1/\sqrt{2})(\chi_2 + \chi_3 + \chi_5 + \chi_6) \quad (\chi_1 + \chi_4) \\
\begin{array}{c}(1/\sqrt{2})(\chi_2 + \chi_3 + \chi_5 + \chi_6)\\[20pt](\chi_1 + \chi_4)\end{array}
\left|\begin{array}{cc} (2X+2) & 2\sqrt{2} \\[20pt] 2\sqrt{2} & (2X+2) \end{array}\right| = 0
\end{array}
$$

$$3.5\text{-}18$$

Using our rules for filling in secular determinant elements, we could have

---

* The form of the group oribtal is important but not the absolute magnitude, since the extent of its contribution to the total MO remains to be determined in the mixing process.

† Note that any multiplication or division of row i and column i of a determinant also multiplies or divides the orbital labeling row and column i.

written this determinant directly, had we been able to predict the desired form. Addition–subtraction of columns and rows diagonalizes the determinant to

$$
\begin{vmatrix}
\begin{bmatrix} (1/\sqrt{2})(x_2 + x_3 + x_5 + x_6) \\ + (x_1 + x_4) \end{bmatrix} & \begin{bmatrix} (1/\sqrt{2})(x_2 + x_3 + x_5 + x_6) \\ - (x_1 + x_4) \end{bmatrix} \\[2em]
(4X + 4 + 4\sqrt{2}) & 0 \\[1em]
0 & (4X + 4 - 4\sqrt{2})
\end{vmatrix} = 0 \qquad 3.5\text{-}19
$$

By this device we have found the proper linear combination of the A_1 symmetry orbitals corresponding to the final (although unnormalized) MOs; these are given as headings of the columns and rows. The corresponding energies result from solution of the separate 1×1 determinants. The normalized MOs are

$$\psi_1 = \tfrac{1}{2}(\chi_1 + \chi_4) + (1/2\sqrt{2})(\chi_2 + \chi_3 + \chi_5 + \chi_6), \quad X = -1 -\sqrt{2} \quad 3.5\text{-}20$$

$$\psi_4 = \tfrac{1}{2}(\chi_1 + \chi_4) - (1/2\sqrt{2})(\chi_2 + \chi_3 + \chi_5 + \chi_6), \quad X = -1 +\sqrt{2} \quad 3.5\text{-}21$$

The two B_2 group orbitals, obtained by application of the B_2 characters 1, -1, -1, 1 to the rows of Table 3.5-6, are found to be $\chi_1 - \chi_4$ and $\chi_2 - \chi_3 - \chi_5 + \chi_6$. The second-order secular determinantal equation mixing these two orbitals is

$$\begin{array}{c} \\ (\chi_2 - \chi_3 - \chi_5 + \chi_6) \\ \\ (\chi_1 - \chi_4) \end{array} \begin{array}{cc} (\chi_2 - \chi_3 - \chi_5 + \chi_6) & (\chi_1 - \chi_4) \\ \left| \begin{array}{cc} (4X - 4) & 4 \\ \\ 4 & (2X - 2) \end{array} \right| \end{array} = 0 \quad 3.5\text{-}22$$

Precisely parallel simplification as followed with 3.5-15 affords the two MOs and their energies as

$$\psi_3 = \tfrac{1}{2}(\chi_1 - \chi_4) + (1/2\sqrt{2})(\chi_2 - \chi_3 - \chi_5 + \chi_6) \quad 3.5\text{-}23$$

with

$$X = 1 - \sqrt{2} \quad 3.5\text{-}24$$

and

$$\psi_6 = \tfrac{1}{2}(\chi_1 - \chi_4) - (1/2\sqrt{2})(\chi_2 - \chi_3 - \chi_5 + \chi_6) \quad 3.5\text{-}25$$

with

$$X = 1 + \sqrt{2} \quad 3.5\text{-}26$$

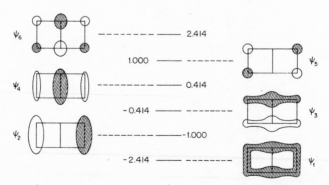

FIG. 3.5-E.   1,4-Dehydrobenzene MOs. The shaded areas represent the positive orbital, the unshaded areas the negative orbital sign above the molecular plane.

The six 1,4-dehydrobenzene MOs can be depicted in qualitative shorthand as in Fig. 3.5-E. Here the shaded portions represent positive orbital signs above the plane of the paper while unshaded portions represent negative signs above this plane. The π energy of 1,4-dehydrobenzene is seen to total $-7.656|\beta|$, giving a resonance energy of $-1.656|\beta|$. Interestingly, this π energy is less than for benzene itself $(-8|\beta|)$. We have here one of a general class of compounds wherein extra overlap leads to destabilization of the molecule.

3.5d Use of Character Tables of Degenerate Groups

Consider species such as cyclopropenyl (Fig. 3.5-F). The approach used in Section 3.1d can be used to afford the MOs and energies. This employed group orbitals symmetric and antisymmetric with respect to σ_{v1}:

$$\psi_1 = (1/\sqrt{3})(\chi_1 + \chi_2 + \chi_3), \qquad X = -2 \qquad 3.5\text{-}27$$

$$\psi_2 = (1/\sqrt{2})(\chi_2 - \chi_3), \qquad X = +1 \qquad 3.5\text{-}28$$

$$\psi_3 = (1/\sqrt{6})(2\chi_1 - \chi_2 - \chi_3), \qquad X = +1 \qquad 3.5\text{-}29$$

However, we might wish to use the formal group theoretical approach delineated in the preceding subsection. There is indeed a group table, C_{3v} (Table 3.5-7), which makes use of all of the symmetry operations of

TABLE 3.5-7
CHARACTERS OF THE C_{3v} GROUP

C_{3v} group	E	$2C_3$	$3\sigma_v$
A_1	1	1	1
A_2	1	1	-1
E	2	-1	0

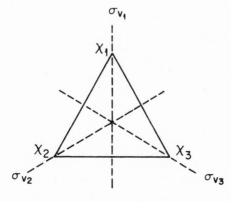

FIG. 3.5-F. Cyclopropenyl and its symmetry operations.

interest* to us in dealing with the cyclopropenyl species. The reader doubt-lessly will note that there are several new features in the C_{3v} table not found in the character tables considered thus far. First of all, there is more than one C_3 operation possible; one can rotate (e.g.) cyclopropenyl by 120° (i.e., C_3) and also by 240° (i.e., $C_3{}^2$). Each of these rotations has the same character for any given irreducible representation; and, rather than list C_3 and $C_3{}^2$ as separate columns we give only one column and head this with the label $2C_3$. In the same way the heading $3\sigma_v$ is shorthand notation indicating that one should really envisage three columns, each with the same characters; these columns correspond to the three vertical reflection operations σ_{v1}, σ_{v2}, and σ_{v3} (cf. 3.5-A for definitions of these operations).

Second, the reader will note that there is an E representation. "E" signifies a (twofold) degenerate representation. The characters of this representation are 2, -1, and 0. Clearly 2 and 0 are not the usual symmetry eigenvalues. In degenerate representations the characters are no longer identical with symmetry eigenvalues, and the reasons can now be discussed.

In looking at the degenerate pair of MOs given in Eqs. 3.5-28 and 3.5-29 ($X = +1$ for both), we note that not all of the C_{3v} group symmetry operators transform these MOs into themselves (eigenvalue then $+1$) or into their negative (eigenvalue then -1). For example, the result of C_3 on ψ_2 is seen† to give $(1/\sqrt{2})(\chi_3 - \chi_1)$. It can be shown that this result, al-though not derived from ψ_2 *alone*, can be expressed as a linear combination of *both* ψ_2 and ψ_3. Thus

$$C_3\psi_2 = (1/\sqrt{2})(\chi_3 - \chi_1) = -\tfrac{1}{2}\psi_2 - (\sqrt{3}/2)\psi_3 \qquad 3.5\text{-}30$$

Similarly

$$C_3\psi_3 = (1/\sqrt{6})(2\chi_2 - \chi_3 - \chi_1) = (\sqrt{3}/2)\psi_2 - \tfrac{1}{2}\psi_3 \qquad 3.5\text{-}31$$

In each case (3.5-30 and 3.5-31) the reader may verify the results by in-spection of Fig. 3.5-F, using the C_3 operator and then by taking the appro-priate combinations of ψ_2 and ψ_3 as given by 3.5-28 and 3.5-29. Further, the reader can demonstrate to his satisfaction that linear combinations of the two degenerate pairs of MOs, ψ_2 and ψ_3, result when other C_{3v} group operations are performed on either MO. Thus in general

$$R\psi_2 = a\psi_2 + b\psi_3 \qquad 3.5\text{-}32$$

and

$$R\psi_3 = c\psi_2 + d\psi_3 \qquad 3.5\text{-}33$$

---

* It does not include σ_h, reflection in the molecular plane, for this operation merely inverts all of the p orbitals and does not aid in finding symmetry orbitals.

† This is obtained by inspection of Fig. 3.5-F and noting the result of C_3 on the in-dividual AOs of ψ_2 as given by 3.5-28.

Given a symmetry operator and a set of orbitals, whether they be atomic orbitals as considered earlier or MOs as ψ_2 and ψ_3 of Eqs. 3.5-32 and 3.5-33, we find that the operator R acting on each orbital will give some fraction of this orbital unchanged plus some additional orbitals. The fraction unchanged for each orbital member of the set is the orbital's contribution to the total character of the set. In the instance of the degenerate set of MOs ψ_2 and ψ_3, the symmetry operator R leaves the fraction a of ψ_2 unchanged while the fraction d of ψ_3 remains unaltered by R. Thus a and d may be considered to be contributions to the total character for operation R of the degenerate set of MOs and $(a + d)$ is the total character. While the individual contributions depend on the precise form of the degenerate MOs, and this has a degree of flexibility,* the total character is independent of the form of the MOs selected.

For the case of the C_3 operation on the MOs ψ_2 and ψ_3 as given in Eqs. 3.5-30 and 3.5-31 we see a and d individually to be $-\frac{1}{2}$ and the total character $(a + d)$ to be -1; this is the character as given in Table 3.5-7 for the E (degenerate) representation and the C_3 operator.

The characters of degenerate groups can be more readily derived by use of matrix methods, but for present purposes attention is focused on practical use of character tables of degenerate groups. Actually, the procedure in dealing with degenerate groups is essentially the same as followed in Section 3.5c for nondegenerate groups.

Assuming that we have not obtained the cyclopropenyl system energies and MOs by use of the single symmetry plane σ_{v1} (cf. Fig. 3.5-F), we proceed to use Rule I of Section 3.5c. The AOs are listed in Table 3.5-8 together with the result of the C_{3v} group operators on these. By inspection of Table 3.5-7 we can see that it will require the characters of the E plus the A_1

---

* Any normalized linear combination of a set of degenerate MOs will be an acceptable MO having the same energy, giving the option of an infinite choice in selecting the form of the first member of such a degenerate pair. However, the second MO is then fixed by the requirement that it be orthogonal to the first member as chosen.

The first statement is readily seen as follows. If ψ_a and ψ_b are an orthonormal set of degenerate MOs of energy E, the energy of any linear combination, say $a\psi_a + b\psi_b$, is given by

$$\int (a\psi_a + b\psi_b)\mathcal{H}(a\psi_a + b\psi_b)\,dv$$

$$= a^2 \int \psi_a \mathcal{H} \psi_a\,dv + 2ab \int \psi_a \mathcal{H} \psi_b\,dv + b^2 \int \psi_b \mathcal{H} \psi_b\,dv$$

$$= a^2 E + 0 + b^2 E = (a^2 + b^2)E,$$

which will afford E if $a^2 + b^2$ is chosen equal to 1. Note that the orthogonality of ψ_a and ψ_b is utilized to set $\int \psi_a \mathcal{H} \psi_b\,dv = 0$.

TABLE 3.5-8

Atomic orbital	Result of group operator					
	E	C_3	$C_3{}^2$	σ_{v1}	σ_{v2}	σ_{v3}
χ_1	χ_1	χ_2	χ_3	χ_1	χ_3	χ_2
χ_2	χ_2	χ_3	χ_1	χ_3	χ_2	χ_1
χ_3	χ_3	χ_1	χ_2	χ_2	χ_1	χ_3
Total character of the reducible representation	3	0	0	1	1	1

representations to give the total character found. More formally, Rule I gives

Number of A_1 symmetry orbitals a_{A1}

$$= \tfrac{1}{6}[1 \cdot 3 + 1 \cdot 0 + 1 \cdot 0 + 1 \cdot 1 + 1 \cdot 1 + 1 \cdot 1] = 1 \qquad 3.5\text{-}34$$

Number of A_2 symmetry orbitals a_{A2}

$$= \tfrac{1}{6}[1 \cdot 3 + 1 \cdot 0 + 1 \cdot 0 - 1 \cdot 1 - 1 \cdot 1 - 1 \cdot 1] = 0 \qquad 3.5\text{-}35$$

Number of E degenerate pairs a_E

$$= \tfrac{1}{6}[2 \cdot 3 - 1 \cdot 0 - 1 \cdot 0 + 0 \cdot 1 + 0 \cdot 1 + 0 \cdot 1] = 1 \qquad 3.5\text{-}36$$

Now we proceed to use the AOs of the rows of Table 3.5-8 and the characters of Table 3.5-7 as coefficients according to Rule II of Section 3.5c. Row 1 of Table 3.5-8 together with the A_1 characters of Table 3.5-7 affords the A_1 orbital $2\chi_1 + 2\chi_2 + 2\chi_3$. Use of rows 2 and 3 merely repeats this as would be anticipated from the prediction by Rule I of only one A_1 orbital. This orbital when normalized is then the same as Eq. 3.5-27.

Despite the prediction of no A_2 orbitals by Rule I, we could attempt to obtain one using the A_2 characters of the C_{3v} table together with the AOs of Table 3.5-8. This is to no avail, as zero results in each case.

Application of the E characters $(2, -1, -1, 0, 0, 0)$ of the C_{3v} table (3.5-7) as coefficients for the AOs of Table 3.5-8 gives us (after dividing by 2) the three group orbitals:

$$\phi_1 = 2\chi_1 - \chi_2 - \chi_3 \qquad 3.5\text{-}37$$

$$\phi_2 = -\chi_1 + 2\chi_2 - \chi_3 \qquad 3.5\text{-}38$$

$$\phi_3 = -\chi_1 - \chi_2 + 2\chi_3 \qquad 3.5\text{-}39$$

We soon see that Rule I, predicting only one pair of degenerate orbitals (E representation), has not led us astray. Of the three group orbitals, only two are independent. ϕ_1 is the first of the two which may be selected; it is just the unnormalized form of the MO ψ_3 given in 3.5-29. $\phi_2 - \phi_3 = 3\chi_2 - 3\chi_3$ is the second of the degenerate pair; this is the unnormalized form of ψ_2 as given in 3.5-28. $\phi_2 + \phi_3 = -2\chi_1 + \chi_2 + \chi_3$ is a repetition of ϕ_1, being just the negative of this. The trick of adding and subtracting two MOs* is a convenient device for obtaining a set of degenerate MOs orthogonal to each other.

In similar fashion the C_{6v} group operations are found sufficient to interconvert every pair of equivalent p orbitals of benzene; and this group, containing two degenerate representations, may be used to afford the solution to the benzene problem (Fig. 3.5-G). The C_{6v} group has 12 operators: the E operator, one C_2 operator, two C_3 operators (C_3 and $C_3{}^2$), two C_6 operators (C_6 and $C_6{}^5$), three reflection operators utilizing planes of symmetry bisecting sides of the benzene ring (the σ_v's), and the three diagonal reflection operators using planes passing through carbon atoms (the σ_d's). The result of these group operators on the six benzene AOs is given in Table 3.5-9. The C_{6v} group characters are given in Table 3.5-10. At the bottom of the

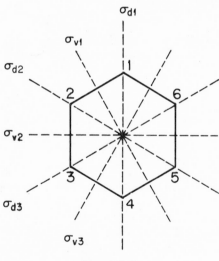

Fig. 3.5-G

* For this device to work the degenerate MOs should either already be individually normalized or should be unnormalized by the same factor as in the example above. Thus in general we see that

$$\int (\psi_1 + \psi_2)(\psi_1 - \psi_2)\, dv = \int \psi_1{}^2\, dv - \int \psi_2{}^2\, dv = 0.$$

and hence $(\psi_1 + \psi_2)$ and $(\psi_1 - \psi_2)$ are indeed orthogonal.

TABLE 3.5-9

EFFECT OF C_{6v} GROUP OPERATORS IN BASIC SET OF BENZENE AOs

AO	Result of group operator											
	E	C_2	C_3	$C_3{}^2$	C_6	$C_6{}^5$	σ_{v1}	σ_{2v}	σ_{3v}	σ_{d1}	σ_{d2}	σ_{d3}
χ_1	χ_1	χ_4	χ_3	χ_5	χ_2	χ_6	χ_2	χ_4	χ_6	χ_1	χ_3	χ_5
χ_2	χ_2	χ_5	χ_4	χ_6	χ_3	χ_1	χ_1	χ_3	χ_5	χ_6	χ_2	χ_4
χ_3	χ_3	χ_6	χ_5	χ_1	χ_4	χ_2	χ_6	χ_2	χ_4	χ_5	χ_1	χ_3
χ_4	χ_4	χ_1	χ_6	χ_2	χ_5	χ_3	χ_5	χ_1	χ_3	χ_4	χ_6	χ_2
χ_5	χ_5	χ_2	χ_1	χ_3	χ_6	χ_4	χ_4	χ_6	χ_2	χ_3	χ_5	χ_1
χ_6	χ_6	χ_3	χ_2	χ_4	χ_1	χ_5	χ_3	χ_5	χ_1	χ_2	χ_4	χ_6
Total character of the reducible representation	6	0	0	0	0	0	0	0	0	2	2	2

TABLE 3.5-10

C_{6v} CHARACTER TABLE

Representation	Symmetry operators					
	E	C_2	$2C_3$	$2C_6$	$3\sigma_v$	$3\sigma_d$
A_1	1	1	1	1	1	1
A_2	1	1	1	1	-1	-1
B_1	1	-1	1	-1	1	-1
B_2	1	-1	1	-1	-1	1
E_1	2	-2	-1	1	0	0
E_2	2	2	-1	-1	0	0
Reducible representation of benzene AOs	6	0	0	0	0	2

group table there are given the characters of the reducible representation of the benzene system of AOs. We must remember that each of the third and fourth columns in actuality represents two columns of characters while the last two columns are each shorthand for three columns. This abbreviation is possible because the characters for any individual representation (e.g., A_2) will be the same for all symmetry operations of the same class (e.g., C_3 and $C_3{}^2$). However, the order of the group (h) is 12; and in using Rules I and II, we must use each column as many times as is indicated by the number preceding the operator heading the column. This is the number of columns which would be written were the group table written out in full.

Application of Rule I gives the number of symmetry orbitals of each symmetry type (irreducible representation):

$$a_{A_1} = \tfrac{1}{12}(6 + 2 + 2 + 2) = 1$$

$$a_{A_2} = \tfrac{1}{12}(6 - 2 - 2 - 2) = 0$$

$$a_{B_1} = \tfrac{1}{12}(6 - 2 - 2 - 2) = 0$$

$$a_{B_2} = \tfrac{1}{12}(6 + 2 + 2 + 2) = 1$$

$$a_{E_1} = \tfrac{1}{12}(12) = 1 \qquad \text{(i.e., one } E_1 \text{ degenerate pair)}$$

$$a_{E_2} = \tfrac{1}{12}(12) = 1 \qquad \text{(i.e., one } E_2 \text{ degenerate pair)}$$

Application of Rule II, using successive rows of Table 3.5-9 together with the characters of the C_{6v} table as coefficients, gives us the eight group orbitals listed below,

$$\phi_{A_1} = 2(\chi_1 + \chi_2 + \chi_3 + \chi_4 + \chi_5 + \chi_6)$$

which after normalization is the MO ψ_1 of page 77.

$$\phi_{B_2} = 2(\chi_1 - \chi_2 + \chi_3 - \chi_4 + \chi_5 - \chi_6)$$

which after normalization is the MO ψ_6 of page 77.

$$\phi_{E_1}' = 2\chi_1 + \chi_2 - \chi_3 - 2\chi_4 - \chi_5 + \chi_6$$

$$\phi_{E_1}'' = \chi_1 + 2\chi_2 + \chi_3 - \chi_4 - 2\chi_5 - \chi_6$$

$$\phi_{E_1}''' = -\chi_1 + \chi_2 + 2\chi_3 + \chi_4 - \chi_5 - 2\chi_6$$

only two of which are independent. ϕ_{E_1}' is, after normalization, the ψ_3 of page 77. The sum and difference of ϕ_{E_1}'' and ϕ_{E_1}''' give unnormalized ψ_2 and ψ_3, the latter being a repetition.

$$\phi_{E_2}' = 2\chi_1 - \chi_2 - \chi_3 + 2\chi_4 - \chi_5 - \chi_6$$

$$\phi_{E_2}'' = \chi_1 - 2\chi_2 + \chi_3 + \chi_4 - 2\chi_5 + \chi_6$$

$$\phi_{E_2}''' = \chi_1 + \chi_2 - 2\chi_3 + \chi_4 + \chi_5 - 2\chi_6$$

of which, similar to the E_1 situation, only two are linearly independent; the sum and difference of ϕ_{E_2}'' and ϕ_{E_2}''' give the MOs ψ_4 and ψ_5 after normalization.

Having obtained the benzene MOs by group theory, we would proceed to find the energies in the usual manner. If more than one group orbital of the same symmetry resulted from Rules I and II, these symmetry orbitals would have to be mixed in the usual manner in a secular determinant and the weighting of these determined by the method of cofactors. In fact, even with degenerate pairs of a given symmetry, in place of orthogonalizing by addition–subtraction, we could have mixed any pair in a 2×2 secular determinant and obtained final MOs and coefficients in the traditional way.

3.5e Use of Group Tables with Complex Characters and Use of Complex Wavefunctions

Thus far we have avoided using group theoretical tables which have complex characters. Among these are the C_n tables. However, these are not difficult to use, and there are instances where they are used to advantage.

For illustration some of the C_n group tables are given in Table 3.5-11.

TABLE 3.5-11
THE C_n GROUP TABLES[a]

C_3		E	C_3	$C_3{}^2$
A		ω^0	ω^0	ω^0
E	E'	ω^0	ω^1	ω^2
	E''	ω^0	ω^{-1}	ω^{-2}

C_4		E	C_4	C_2	$C_4{}^3$
A		ω^0	ω^0	ω^0	ω^0
E_1	E_1'	ω^0	ω^1	ω^2	ω^3
	E_1''	ω^0	ω^{-1}	ω^{-2}	ω^{-3}
B		ω^0	ω^2	ω^4	ω^6

C_5		E	C_5	$C_5{}^2$	$C_5{}^3$	$C_5{}^4$
A		ω^0	ω^0	ω^0	ω^0	ω^0
E_1	E_1'	ω^0	ω^1	ω^2	ω^3	ω^4
	E_1''	ω^0	ω^{-1}	ω^{-2}	ω^{-3}	ω^{-4}
E_2	E_2'	ω^0	ω^2	ω^4	ω^6	ω^8
	E_2''	ω^0	ω^{-2}	ω^{-4}	ω^{-6}	ω^{-8}

C_6		E	C_6	C_3	C_2	$C_3{}^2$	$C_6{}^5$
A		ω^0	ω^0	ω^0	ω^0	ω^0	ω^0
E_1	E_1'	ω^0	ω^1	ω^2	ω^3	ω^4	ω^5
	E_1''	ω^0	ω^{-1}	ω^{-2}	ω^{-3}	ω^{-4}	ω^{-5}
E_2	E_2'	ω^0	ω^2	ω^4	ω^6	ω^8	ω^{10}
	E_2''	ω^0	ω^{-2}	ω^{-4}	ω^{-6}	ω^{-8}	ω^{-10}
B		ω^0	ω^3	ω^6	ω^9	ω^{12}	ω^{15}

[a] In C_3, $\omega = e^{2\pi i/3}$ and $C_3{}^3 = E$. In C_4, $\omega = e^{2\pi i/4}$ and $C_4{}^4 = E$. In C_5, $\omega = e^{2\pi i/5}$ and $C_5{}^5 = E$. In C_6, $\omega = e^{2\pi i/6}$ and $C_6{}^6 = E$.

The characters are given as functions of a variable ω. It is seen that the tables have a simple pattern. Thus, the group operators correspond to rotations by multiples of $1/n$ in each $\mathbf{C}_n$ group table. The first operator in each case then is a zero-degree rotation (i.e., the E operator) corresponding to $C_n{}^0$. The next operator is $C_n{}^1$, followed by $C_n{}^2$, $C_n{}^3$, etc., until all possible rotations are considered. In cases where C_n to some power is more simply expressed, such as $C_6{}^2 = C_3$ in the $\mathbf{C}_6$ table, this is done. The symmetric representation is just a row vector of ones but to emphasize the pattern, in each table this is given as $[\omega^0 \ \omega^0 \ \omega^0 \ \omega^0 \cdots \omega^0]$.

Next we note that the variable ω is defined as a function of an angle θ and that θ is just the angle of rotation corresponding to the group operator C_n. Note

$$\omega = e^{i\theta} \qquad\qquad 3.5\text{-}40$$

where

$$\theta = (2\pi/n) \qquad\qquad 3.5\text{-}41\text{a}$$

and thus

$$\omega = e^{2\pi i/n} \qquad\qquad 3.5\text{-}41\text{b}$$

However, we can express ω as a linear combination of real and imaginary parts as in the following:

$$\omega = \cos\theta + i\sin\theta \qquad\qquad 3.5\text{-}42\text{a}$$

$$\omega = \cos(2\pi/n) + i\sin(2\pi/n) \qquad\qquad 3.5\text{-}42\text{b}$$

It can be seen that doubling the angle of rotation θ corresponds to squaring ω; i.e., $\omega^2 = e^{i2\theta}$. Tripling the angle θ likewise corresponds to cubing ω. In general, raising ω to any power (e.g., r) is equivalent to multiplying the angle by that power (i.e., giving an angle $r\theta$). Thus,

$$\omega^r = \cos(r\theta) + i\sin(r\theta) = \cos(2\pi r/n) + i\sin(2\pi r/n) \qquad 3.5\text{-}43$$

Two further generalities are of use. First it can be seen from Eq. 3.5-43 that

$$\omega^n = 1 \qquad\qquad 3.5\text{-}44$$

(i.e., here $r = n$). This results since the cosine term in Eq. 3.5-43 becomes unity while the sine term becomes zero. The other generality is that

$$\omega^1 + \omega^2 + \omega^3 + \omega^4 + \cdots + \omega^n = 0 \qquad\qquad 3.5\text{-}45$$

This is seen by setting S equal to this sum as in 3.5-46 and subtracting the quantity ωS as in 3.5-47 to give 3.5-48:

$$S = \omega^1 + \omega^2 + \omega^3 + \omega^4 + \cdots + \omega^n \qquad\qquad 3.5\text{-}46$$

$$\omega S = \omega^2 + \omega^3 + \omega^4 + \cdots + \omega^{n+1} \qquad\qquad 3.5\text{-}47$$

$$S(1 - \omega) = \omega^1 - \omega^{n+1} = \omega(1 - \omega^n) \qquad\qquad 3.5\text{-}48$$

From Eq. 3.5-44, we know that $\omega^n = 1$, and thus the right-hand side of 3.5-48 vanishes. Since we know that ω has a variety of values and is not generally unity, the left-hand side of 3.5-48 can be zero only if $S = 0$, thus proving 3.5-45.

In inspecting the group tables we find that all of these begin with a symmetric (i.e., A) representation followed by a series of degenerate representations. The even-dimensioned groups also have a B; the odd groups do not. Also, we see that the degenerate representations are given explicitly for both members of each degenerate pair. Further examination shows that in each representation the characters of the successive operators differ by an increment in the exponent. Beyond this, in proceeding from the symmetric representation through the first members of each degenerate representation and onward to the B representation, where it exists, this increment in the powers increases by one as each new representation is encountered. Thus the A representation characters are functions of an angle which does not increase at all in proceeding from operator to operator; the E_1' representation characters are functions of an angle which increases by $2\pi/n$ in going from one character to the next of the representation; the E_2' increment is $4\pi/n$; the E_3' increment is $6\pi/n$; and so on.

For complex representations the characters of the second member of each degenerate pair are the complex conjugate of the corresponding characters of the first member; that is, each pair of characters from the same representation and the same rotation gives unity when multiplied together.

Also, for complex groups we need a more general definition of orthonormality as can be seen from multiplication of the two vector members of a degenerate pair. This does not give zero but rather gives h, the order of the group. Similarly, the result of taking the sums of squares of the characters of a single member of a degenerate pair is zero rather than h.

Thus the test for normalization is taking the sum of the products of the characters multiplied by their complex conjugates which should equal the order of the group. And, the test for orthogonality is that the products of the characters of one member multiplied by the complex conjugates of the corresponding characters of the other member should sum to give zero. These two tests are given in Eqs. 3.5-49 and 3.5-50.

$$(1/h) \sum_R X_{rR}{}^*X_{sR} = 0 \qquad \text{(summation over all group operators, } R) \qquad 3.5\text{-}49$$

$$(1/h) \sum_R X_{rR}{}^*X_{rR} = 1 \qquad\qquad\qquad\qquad\qquad 3.5\text{-}50$$

One example of the use of such complex group tables is the cyclopropenyl problem (note Fig. 3.5-H). If we apply Rule I to the reducible representation derived in Fig. 3.5-H, we find that one of each of the three irreducible

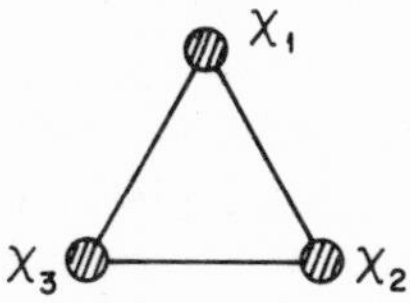

	E	C_3	$C_3{}^2$
	χ_1	χ_2	χ_3
	χ_2	χ_3	χ_1
	χ_3	χ_1	χ_2
$\Gamma_{\text{red}} =$	$[3$	0	$0]$

FIG. 3.5-H.   The cyclopropenyl basis set and its reducible representation.

representations that follows is present:

$$a_A = (1/h)\,\Gamma_{\text{red}}{}^*\,\tilde{\Gamma}_A = \tfrac{1}{3}[3 \quad 0 \quad 0]\begin{bmatrix} \omega^0 \\ \omega^0 \\ \omega^0 \end{bmatrix} = 1 \qquad\qquad 3.5\text{-}51$$

$$a_{E'} = (1/h)\,\Gamma_{\text{red}}{}^*\,\tilde{\Gamma}_{E'} = \tfrac{1}{3}[3 \quad 0 \quad 0]\begin{bmatrix} \omega^0 \\ \omega^1 \\ \omega^2 \end{bmatrix} = 1 \qquad\qquad 3.5\text{-}52$$

$$a_{E''} = (1/h)\,\Gamma_{\text{red}}{}^*\,\tilde{\Gamma}_{E''} = \tfrac{1}{3}[3 \quad 0 \quad 0]\begin{bmatrix} \omega^0 \\ \omega^{-1} \\ \omega^{-2} \end{bmatrix} = 1 \qquad\qquad 3.5\text{-}53$$

Now we proceed to apply Rule II to obtain the group orbitals. For convenience we select the first row of the transformed basis orbitals in Fig. 3.5-H, and we use the characters in the C_3 group table as LCAO-MOs. coefficients. This gives us three orbitals of different symmetry, and these are therefore eigenfunctions directly. They are

$$\theta_1 = \omega^0\chi_1 + \omega^0\chi_2 + \omega^0\chi_3 \qquad\qquad 3.5\text{-}54a$$

$$\theta_2 = \omega^0\chi_1 + \omega^1\chi_2 + \omega^2\chi_3 \qquad\qquad 3.5\text{-}54b$$

$$\theta_3 = \omega^0\chi_1 + \omega^{-1}\chi_2 + \omega^{-2}\chi_3 \qquad\qquad 3.5\text{-}54c$$

These are still unnormalized, as can be shown, and need a factor of $1/\sqrt{3}$. Although ω^0 is unity it is retained to show the equivalence of all three cyclopropenyl centers. Each AO is weighted similarly by a function of an angle (i.e., some power of ω), and the angle changes by a constant increment in proceeding from AO to AO. Thus the successive coefficients in one MO differ only by a phase factor. The phase factor has opposite sign for degenerate MOs and these "rotate" in opposite directions. Also, we can multiply all MOs by any power of ω and still have an acceptable set of MOs.

In using complex wavefunctions such as 3.5-54a,b,c to obtain electron densities, bond orders, MO energies, and so on, we must use a more general form of the appropriate integrals in which one of the two wavefunctions is written as the complex conjugate. With real MOs this assumes the simple form we have already been using. For example, to get energy, we proceed as follows (MO 2 is used as an example):

$$E_2 = \int \theta_2{}^* \mathfrak{K} \theta_2 \, d\tau$$

$$= \tfrac{1}{3} \int (\chi_1 + \omega^{-1}\chi_2 + \omega^{-2}\chi_3) \mathfrak{K} (\chi_1 + \omega\chi_2 + \omega^2\chi_3) \, d\tau$$

$$= \tfrac{1}{3}[3\alpha + 3(\omega^{-1} + \omega^1)\beta] = \alpha + 2\cos(2\pi/3)\beta$$

$$= \alpha - \beta \quad \text{i.e.,} \quad X = +1 \qquad\qquad 3.5\text{-}55$$

In the case of electron densities and bond orders, one can obtain meaningful results from one member of a degenerate pair, in contrast to the situation with the real form of the MOs where one needs to take a degenerate pair of MOs together in order to obtain results which are not dependent on what linear combination is selected for use.

One other point needs to be made about the complex wavefunctions in Eqs. 3.5-54b and 3.5-54c. This point is illustrated by taking linear combinations of the degenerate MOs θ_2 and θ_3; here we will take the sum and difference. We obtain

$$\psi_+ = (\theta_2 + \theta_3) = [2\omega^0\chi_1 + (\omega^1 + \omega^{-1})\chi_2 + (\omega^2 + \omega^{-2})\chi_3] \quad 3.5\text{-}56a$$

$$\psi_- = (\theta_2 - \theta_3) = [(\omega^1 - \omega^{-1})\chi_2 + (\omega^2 - \omega^{-2})\chi_3] \qquad 3.5\text{-}56b$$

But reference to Eq. 3.5-43 allows us to express ω^1, ω^2, ω^{-1}, and ω^{-2} explicitly. We find that $(\omega^1 + \omega^{-1}) = 2\cos(2\pi/3) = -1$. Similarly, $(\omega^2 + \omega^{-2}) = -1$. This allows us to express 3.5-56 as

$$\psi_+ = 2\chi_1 - \chi_2 - \chi_3 \qquad\qquad 3.5\text{-}57a$$

which is still unnormalized but is seen to be the usual real form of one of the two degenerate MOs of cyclopropenyl. In similar fashion we obtain $(\omega^1 - \omega^{-1}) = 2i\sin(2\pi/3)$ and $(\omega^2 - \omega^{-2}) = -2i\sin(2\pi/3)$. Thus the two coefficients of χ_2 and χ_3 have equal magnitudes but opposite signs. Since the MOs are not yet normalized we might just as well use $+1$ and -1 as coefficients and write the resulting MO as

$$\psi_- = \chi_2 - \chi_3 \qquad\qquad 3.5\text{-}57b$$

which is seen to be the usual form for the real, unnormalized second degenerate member for cyclopropenyl.

In general, one can convert the complex MOs into real forms by just adding and subtracting the degenerate pairs in this way followed by normalization.

3.6 Complex Characters in Deriving the Hückel and Möbius Solutions

Previously we introduced the Hückel and Möbius formulas without proof. However, now that we have considered complex characters it is of interest to reconsider the problem.

We begin with the matrix formulation of the eigenvalue problem as

$$\mathbf{Hc} = \mathbf{Xc} \qquad \text{or} \qquad [\mathbf{H} - \mathbf{X}]\mathbf{c} = 0 \qquad\qquad 3.6\text{-}1a,b$$

$$[\mathbf{X} - \mathbf{H}]\mathbf{c} = 0 \qquad\qquad 3.6\text{-}1c$$

The form in 3.6-1c is identical to that which we have formerly used, with a determinant of X's and zeros, in dealing with all types of delocalized systems. We now rewrite this explicitly in Eq. 3.6-2 with a trial vector for $\mathbf{c}$. The plus sign for the corner elements is used for Hückel systems and the minus sign for Möbius ones. For simplicity a 5×5 secular matrix is used but the result can be thought of generally.

$$\begin{bmatrix} X & 1 & 0 & 0 & \pm 1 \\ 1 & X & 1 & 0 & 0 \\ 0 & 1 & X & 1 & 0 \\ 0 & 0 & 1 & X & 1 \\ \pm 1 & 0 & 0 & 1 & X \end{bmatrix} \begin{bmatrix} \nu^0 \\ \nu^1 \\ \nu^2 \\ \nu^3 \\ \nu^4 \end{bmatrix} = 0 \qquad\qquad 3.6\text{-}2$$

The vector resulting on performing the multiplication shown in Eq. 3.6-2 is zero, and thus each element must be zero. This vector is given in the following equation:

$$
\begin{bmatrix}
\nu^0 X + \nu^1 & \pm \nu^4 \\
\nu^0 & + \nu^1 X + \nu^2 \\
\nu^1 & + \nu^2 X + \nu^3 \\
\nu^2 & + \nu^3 X + \nu^4 \\
\pm \nu^0 & + \nu^3 & + \nu^4 X
\end{bmatrix}
=
\begin{bmatrix}
(X + \nu^1 + \nu^{-1})\nu^0 \\
(X + \nu^1 + \nu^{-1})\nu^1 \\
(X + \nu^1 + \nu^{-1})\nu^2 \\
(X + \nu^1 + \nu^{-1})\nu^3 \\
(X + \nu^1 + \nu^{-1})\nu^4
\end{bmatrix}
= 0 \qquad \text{3.6-3}
$$

In setting up Eq. 3.6-2 we have made use of the implicit assumption that the molecule will conform to a C_n type of group and that we can therefore use a C_n type of representation for the **c** vector. Here, if ν is taken as some power of ω, it can be seen that the **c** vector is indeed a generalized C_n irreducible representation. We keep the exact form of ν undefined for the time being except that, in simplifying the vector obtained (note Eq. 3.6-3), we do assume that

$$
\nu^5 = \pm 1 \qquad \text{3.6-4}
$$

where the plus sign is used for Hückel systems and the minus sign for Möbius systems. This has allowed us to deal with the $\pm \nu^4$ and $\pm \nu^0$ terms in Eq. 3.6-3.

Looking at the resulting vector in this equation, we recognize that the ν^0, ν^1, ν^2, ν^3, and ν^4 terms are nonzero and thus the vector will vanish only if

$$
X = -\nu^1 - \nu^{-1} \qquad \text{3.6-5}
$$

for both Hückel and Möbius systems.

For Hückel systems, where $\nu^5 = +1$, a solution is obtained if we take

$$
\nu = \exp[(ik/n)2\pi] \qquad \text{3.6-6}
$$

where $n = 5$ presently for the five-orbital system. That this is a solution to Eq. 3.6-4 taken with a plus sign is readily seen by considering the trigonometric form (note $e^{i\theta} = \cos\theta + i\sin\theta$) of a complex variable. Also, using this form, we can rewrite Eq. 3.6-5 as

$$
X = -\exp[(ik/n)2\pi] - \exp[-(ik/n)2\pi] = -2\cos(2k\pi/n) \qquad \text{3.6-7}
$$

which is the Hückel formula. Here $k = 0, 1, 2, \ldots, n - 1$ and is the MO number.

In the case of the Möbius system, where ν^5 must equal -1, we can select

ν as in Eq. 3.6-8a; however, a more general solution is seen in Eq. 3.6-8b:

$$\nu = \exp(i\pi/n) \qquad\qquad 3.6\text{-}8a$$

$$\nu = \exp[(2k + 1)i\pi/n] \qquad\qquad 3.6\text{-}8b$$

Using the form in 3.6-8b, we can evaluate the energy expression in Eq. 3.6-5 for Möbius systems. Thus

$$X = -\exp[(2k + 1)i\pi/n] - \exp[-(2k + 1)i\pi/n] = -2\cos(2k + 1)\pi/n$$

$$3.6\text{-}9$$

which is the general formula for Möbius systems. Again k is the MO number.

For both Hückel and Möbius systems there is another result from our efforts above. Thus, having obtained a general expression for these eigenvalues, we know that the vector $[\nu^0\ \nu^1\ \nu^2\ \nu^3 \cdots \nu^n]$ used in Eq. 3.6-2 is the eigenvector **c** and its elements thus give the LCAO-MO coefficients. In the Hückel case, these come from the C_n group tables, and in the Möbius case a similar set of characters applies except with a different definition of ν. We can readily convince ourselves that the complex conjugate vector $[\nu^0\ \nu^{-1}\ \nu^{-2}\ \nu^{-3} \cdots \nu^{-n}]$ applies to the second degenerate member in cases of degenerate pairs. Also we could use sums and differences of the two vectors. Thus the LCAO-MO coefficients in the case of the degenerate molecular orbitals are:

$$c_{rk}{}^+ = [1/(2n)^{1/2}](\nu^r + \nu^{-r}) = (2/n)^{1/2}\cos(2kr\pi/n) \qquad 3.6\text{-}10a$$

$$c_{rk}{}^- = [1/(2n)^{1/2}](\nu^r - \nu^{-r}) = (2/n)^{1/2}\sin(2kr\pi/n) \qquad 3.6\text{-}10b$$

In the case of 3.6-10b, we have dropped a factor of i which was present, since it is a constant multiplier. The trigonometric form for the LCAO-MO coefficients is convenient and common for Hückel cyclic polyenes.

For the nondegenerate Hückel bonding MO we have more simply just

$$c_r = 1/\sqrt{n} \qquad\qquad 3.6\text{-}10c$$

In the case of Möbius systems our treatment differs only in the value of ν employed. Here we obtain

$$c_{rk}{}^+ = (1/\sqrt{2})(\nu^r + \nu^{-r}) = (2/n)^{1/2}\cos(2k + 1)r\pi/n \qquad 3.6\text{-}11a$$

$$c_{rk}{}^- = (1/\sqrt{2})(\nu^r - \nu^{-r}) = (2/n)^{1/2}\sin(2k + 1)r\pi/n \qquad 3.6\text{-}11b$$

for the degenerate pairs of MOs. If there is a nondegenerate antibonding MO, its LCAO-MO coefficients are given by

$$c_r = 1/\sqrt{n}(1)^{-r} \qquad\qquad 3.6\text{-}12$$

Thus, our use of the **c** vector chosen initially has been justified. In the

Hückel case the C_n group table has proved exactly applicable with $\nu = \omega^{k-1}$. For the Möbius case, a C_n group table with special requirements is seen to apply. Here $\omega^{(k-1)n} = \nu^n = -1$, where k again is the MO number.

3.7 Use of Moiety Eigenfunctions in Construction of Molecular Orbitals

Hitherto we have been restricted to symmetry in the simplification of secular determinants. In using symmetry, we assume we know the relative weighting of basis orbitals which are equivalently located in a molecule. However, there are other instances where basis orbitals can be known in advance of calculation to have certain relative weightings. Thus, a group of three orbitals in a linear array is an allyl-like moiety and the final MOs obtained will have allyl-like relative weightings of these basis orbitals as long as the only other moieties in the molecule are also allyl-like and interact in a symmetrical fashion.

Thus, we might consider the layered compound shown in Fig. 3.7-A. Such compounds have been synthesized and have their rings held together with methylene bridges; however, we will ignore the methylene bridges and consider only the p-orbital system. We may deal with this molecule as a sixfold set of linear allyl-like arrays, each consisting of three colinear p orbitals. For example, p orbitals $1a$, $1b$, and $1c$ make up one such array.

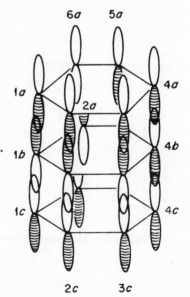

FIG. 3.7-A.   Orbitals for triple-layered compound.

TABLE 3.7-1
IRREDUCIBLE REPRESEN-
TATIONS FOR ALLYL-LIKE
MOIETIES[a]

	R_1	R_2	R_3
Γ_1	$\sqrt{2}$	0	$-\sqrt{2}$
Γ_2	1	$\sqrt{2}$	1
Γ_3	1	$-\sqrt{2}$	1

[a] Normalized to a sum of squares of 4.

Alternatively, we may consider the molecule made up of benzene-like MOs for each of the three layers and then look at the problem of admixing of these MOs.

If we take the first approach, we note that each of the six colinear allyl-like arrays has three MOs. The weighting of allyl MO coefficients will be those in Table 3.7-1. It is now of interest to mix the three allyl-like MOs derived from atoms $1a$, $1b$, and $1c$ with the other five sets making up the triple-layered compound. It is quickly found that allyl orbitals of different representations do not admix. Thus, for example, the lowest energy MO of moiety 1 (i.e., corresponding to representation Γ_2) does not admix with the nonbonding MO of moiety 2 (here corresponding to representation Γ_1). Thus, all we need to do is to admix the six bonding MOs of the six allyl-like colinear arrays as if we were doing the benzene problem, to mix the six nonbonding arrays in the same way, and finally to mix the six antibonding arrays. In these mixing processes one can use either the sixfold symmetry of benzene or instead the two planes of symmetry (i.e., as in C_{2v}). Thus each of the sixth-order determinants will break up into two 2×2's and two 1×1's. Below we give the original sixth-order determinant derived from the bonding allyl MOs and leave it to the reader to complete the treatment.

$$
\begin{array}{c|cccccc}
 & \phi_1 & \phi_2 & \phi_3 & \phi_4 & \phi_5 & \phi_6 \\
\hline
\phi_1 & 4X + 4\sqrt{2}\epsilon & 4 & 0 & 0 & 0 & 4 \\
\phi_2 & 4 & 4X + 4\sqrt{2}\epsilon & 4 & 0 & 0 & 0 \\
\phi_3 & 0 & 4 & 4X + 4\sqrt{2}\epsilon & 4 & 0 & 0 \\
\phi_4 & 0 & 0 & 4 & 4X + 4\sqrt{2}\epsilon & 4 & 0 \\
\phi_5 & 0 & 0 & 0 & 4 & 4X + 4\sqrt{2}\epsilon & 4 \\
\phi_6 & 4 & 0 & 0 & 0 & 4 & 4X + 4\sqrt{2}\epsilon
\end{array} = 0
$$

The diagonal elements of the 6×6 result from such interactions as $(\chi_{1a} + \sqrt{2}\chi_{1b} + \chi_{1c})$ with itself (i.e., ϕ_1 with ϕ_1) to afford 4 squared terms and $4\sqrt{2}$ cross-product terms. In this case the cross-product terms are of the type $\chi_{1a}\chi_{1b}$ and thus each contributes ϵ which depends on the amount of coaxial p–p overlap. The off-diagonal elements result from interactions of the type $(\chi_{1a} + \sqrt{2}\chi_{1b} + \chi_{1c})$ with $(\chi_{2a} + \sqrt{2}\chi_{2b} + \chi_{2c})$, and it is seen that there are four cross-product contributions of the normal parallel p-orbital overlap type.

The second approach involves taking three MOs at a time, one from each benzene moiety and all three of the same symmetry in C_{2v}. These three mix only with one another but not with the other benzene MOs of different symmetry. It can be seen then that each of the six benzene MOs will be split in an "allyl-like fashion" by $-\sqrt{2}\epsilon$, 0, and $+\sqrt{2}\epsilon$. With either approach the 18 resulting MOs become $-2 - \sqrt{2}\epsilon$, -2, $-2 + \sqrt{2}\epsilon$, $\underline{-1 - \sqrt{2}\epsilon}$, $\underline{-1}$, $\underline{-1 + \sqrt{2}\epsilon}$, $\underline{1 - \sqrt{2}\epsilon}$, $\underline{+1}$, $\underline{+1 + \sqrt{2}\epsilon}$, $+2 - \sqrt{2}\epsilon$, $+2$, $+2 + \sqrt{2}\epsilon$. The underlined MOs are degenerate pairs.

One final point of interest is that if we populate the bonding MOs, that is, with the 18 electrons of the three benzene rings, we obtain a π energy which is that of isolated benzene rings and the total π energy does not contain any ϵ terms, meaning that the π energy is independent of distance between rings. Only if the rings are compressed to the point where ϵ exceeds $1/\sqrt{2}$ is stabilization derived. Thus, like barrelene, we have a molecule whose MOs are not those of the isolated component parts, and yet the electrons are delocalized. However, there is no accompanying delocalization energy.

This situation is common to like moieties approaching one another symmetrically. Hence, the method is very general and can be applied to a variety of systems. Thus, the 1,4-dehydrobenzene problem on page 95 could be done more simply by taking plus and minus combinations of the corresponding pairs of the three allyl molecular orbitals. Alternatively, but less simply, one could solve the problem by taking the ethylenic MOs in "allyl-like" linear combinations. First one would use the bonding ethylene MOs and then the antibonding ethylene MOs. Among the problems that follow is included the solution of the barrelene problem by using three ethylenic MO systems, which is still another use of the method.

Problems

1. Using symmetry do all problems at the end of Chapter 1 where symmetry allows direct formulation of the secular determinant.

2. Devise a way to use the circle device to solve the barrelene problem for all six MO energies. Be prepared to justify the application of the monocyclic device to this six-orbital, non-monocyclic problem.

3. Use symmetry to do the naphthalene problem.

4. Use symmetry to do the 7-norbornadienyl problem, and decide whether

the cation, radical, or carbanion is the stable species. Take 2–7 overlap to be ϵ and take the other transannular overlap for simplicity to be ϵ as well. Neglect 2–5 and 3–6 overlap.

5. Redo Problem 4 without carbon-7 having a p orbital (i.e., the norbornadiene problem).

   (a) Which MOs are common to Problem 4 and why?

   (b) Is the delocalization energy a function of ϵ?

6. Get the MO energies for barrelene. Use a basis set with plus lobes aiming clockwise and the bridges labeled 1,2; 3,4; and 5,6. Use 1 for the intrabridge overlap and ϵ for interbridge overlap if plus–plus.

7. If you premultiply the ethylene secular determinant by the row vector $[1 \;\; -1]$ and postmultiply by the column vector $[1 \;\; -1]$, what operation have you performed? What value of X does this give? Try premultiplying by the matrix

$$\begin{bmatrix} 1 & -1 \\ 1 & 1 \end{bmatrix}$$

and postmultiplying by the matrix

$$\begin{bmatrix} 1 & 1 \\ -1 & 1 \end{bmatrix}$$

What does this do, which should be familiar. These are called similarity transformations.

8. Having used one plane of symmetry for solution of a trigonal problem as cyclopropenyl, why can you not proceed to further simplify using a second plane of symmetry (i.e., as a plane going through χ_2 after using one going through χ_1)?

9. Get one solution (i.e., one eigenvalue) for a generalized secular determinant for a cyclic polygon (e.g., benzene) by addition of the elements of rows 2 through n (here 6) to row 1. Note this is an allowed operation. How does this help? What MO energy is obtained?

10. Show that MOs 1 and 2 of allyl are truly orthogonal. Use the definition of orthogonality as a starting point. That is, use the overlap integral between MOs ψ_1 and ψ_2 and the knowledge of the LCAO-MO coefficients for allyl.

11. Use the integrated form of the Schrödinger equation and the LCAO expression for MO 1 of allyl to obtain the energy of MO 1.

12. Premultiply the secular determinant for allyl by

$$\begin{bmatrix} 1 & 0 & 1 \\ 0 & 1 & 0 \\ -1 & 0 & 1 \end{bmatrix}$$

and postmultiply by

$$\begin{bmatrix} 1 & 0 & -1 \\ 0 & 1 & 0 \\ 1 & 0 & 1 \end{bmatrix}.$$

The allyl determinant should have the AOs in the order χ_1, χ_2, χ_3. What operation have you performed which is familiar? Mathematically, you have carried out a similarity transformation except for a scalar "fudge factor."

13. Using two planes of symmetry, set up the separate group orbitals and secular determinants for the naphthalene problem. Use a numbering system beginning with 1 at the junction position. Solve for the eigenvalues and those eigenfunctions which do not require the method of cofactors. Which MOs have you seen before in other molecules?

14. Use symmetry to solve the following problems. Get the eigenvalues for these molecules:

(a)   (b)   (c)

(d) In the problem above, one might have started with MOs for known moieties rather than with atomic orbitals as a basis set. For example, in (b) you might begin with the MOs for the ethylenic moieties. Try this and see whether or not you can predict in advance which MOs from the left ring will interact with which MOs from the right ring? If so, in what general situation will there be interaction?

15. Use the simplified method of setting up determinants directly to solve the barrelene problem for any one of the symmetry types. Use two perpendicular planes. Then list the other symmetry orbitals in groups; however, these need not be solved explicitly.

16. What happens when the p orbital of 7-norbornadienyl is brought in from infinity to the diene moiety? That is, which MOs of norbornadiene itself interact with the p orbital? What are the final MOs?

17. Use three planes of symmetry to solve the paracyclophane problem. Ignore the $(CH_2)_n$ bridges and assume the planarity of the rings. Use an overlap of $\epsilon = S/S_0$. Number the two rings $1a, 2a, 3a, \ldots, 6a$ and $1b, 2b, \ldots,$ $6b$. Take one plane through atom $1a$. Use the D_{2h} group table and set up the subdeterminants directly. Is there appreciable delocalization energy in this molecule beyond that of the two benzene rings? How much?

18. (a) Given a reducible vector $\mathbf{V}_r = 2\mathbf{V}_{A_1} + \mathbf{V}_{B_1} + \mathbf{V}_{B_2}$, multiply by the irreducible vector $\mathbf{V}_{A_1}$ and divide by h, the order of the vector. What is the answer and its significance?

(b) Generalize this into a proof of Rule I. [*Hint*: Start with $\mathbf{V}_r = a_1\mathbf{V}_1 + a_2\mathbf{V}_2 + a_3\mathbf{V}_3 + \cdots + a_n\mathbf{V}_n.$]

19. (a) Take allene as in the accompanying figure, where there is a modified Newman projection diagram. χ_1 is at atom 1, χ_2 and χ_3 at atom 2, and χ_4 at atom 3. χ_2 and χ_3 are kept 45° from horizontal and are not turned. Now consider the 90° twisting of the terminal methylenes, increasing $\theta_{12} = \theta_{34}$ the angle between the first two AOs and between the last two to 90° each. Use the relationship $(\beta_{12}/\beta_0) = \cos\theta_{12}$ as giving the form of resonance integrals between two twisted orbitals. Set up two 2×2 determinants of symmetry orbitals using the C_2 axis. Now solve for energy as a function of angle of twist (i.e., $\theta = 0 \to 90°$). Use this to draw a correlation

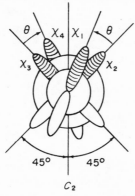

diagram. Is the noncrossing rule violated? How do you explain the result? [*Note*: This requires some thinking!]

(b) What result would you get if you picked the central orbitals as horizontal and vertical instead of diagonal?

(c) Consider the molecule as set up in 19*a* at $\theta = 45°$ and the molecule as set up in 19*b* at initial geometry. Which of the systems is Hückel and which is Möbius? What then is predicted about the MO array at these geometries? Is this what you actually found?

20. Use the pictured unit as a basis. This consists of two *p* orbitals aimed

at one another colinearly. Obtain bonding and antibonding molecular orbitals for the system. Use these resulting molecular orbitals (six bonding and six antibonding such bases are needed) to solve the paracyclophane problem in a fashion reminiscent of the benzene problem.

21. In a similar way, do the barrelene problems by using the MOs of twist-dihydrotrimethylenemethane as bases. Why do you not have to mix

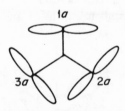

all six MOs together to solve the problem (i.e., all three from the front face of barrelene and all three from the rear face)?

22. Do the paracyclophane problem using symmetry to simplify. To what extent is there stabilization in bringing two aromatic rings together in this fashion?

23. What happens as two cyclobutadiene molecules approach one another face to face with *p* orbitals approaching coaxially? Take each intermolecular overlap as ϵ for each pair of atomic orbitals. Do the problem by using pairs of final cyclobutadiene MOs. Give the final MO energies as a function of ϵ but do not bother to get final eigenfunctions. Do justify at the end your mixing only certain MOs and show your work explicitly, indicating which MOs are mixed and what the secular determinant elements are.

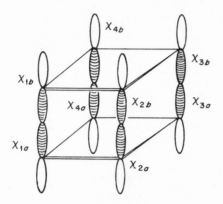

24. Given the representation $[4\ 0\ -2\ -2]$ in the C_2 group, decide how many times the B_2 representation occurs in this. Show how you obtained your answer.

25. Do the 1,3-dehydrocyclobutadiene problem using two planes of symmetry and setting up the secular determinants of each symmetry type separately. Do your work explicitly to obtain the eigenvalues. Label each

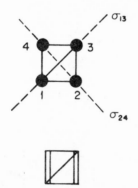

determinant with its symmetry and with the basis orbitals used. Give the final eigenvalues, and only those eigenfunctions which can be obtained without the use of cofactors. (Note overlap 1–3 and assume this is the normal vicinal overlap.)

Suggested Reading

F. A. Cotton, "Chemical Applications of Group Theory." (Wiley Interscience), New York, 1963.

H. Eyring, J. Walter, and G. E. Kimball, "Quantum Chemistry." Wiley, New York, 1958, Chap. 10, Appendix VII.

Chapter 4

EXTENSIONS, MODIFICATIONS, AND APPLICATIONS OF THE HÜCKEL APPROACH

This chapter deals with calculations for molecules with heteroatoms, the means of including overlap into calculations, methods of constructing and using hybrid orbitals and orbitals with unusual orientation, the matrix formulation of the eigenvalue problem, special properties and uses of non-bonding molecular orbital coefficients, and construction of correlation diagrams.

4.1 Calculations for Molecules Containing Heteroatoms

As has been shown by Pauling and Wheland,[1] the secular determinant for a molecule containing a heteroatom (e.g., oxygen or nitrogen) may be written in the usual fashion except that the diagonal element corresponding to the heteroorbital column and row becomes $(X + \delta)$ rather than the usual X. Here δ is a measure of the electronegativity of the heteroatomic orbital. The more positive values correspond to a more electronegative orbital, and conversely. That δ represents an increment in the Coulomb integrals, which occur along the diagonal, will be shown below.

A wide assortment of different values of δ have been used for each element; Table 4.1-1 gives a compilation of values recommended by Streitwieser[2] for oxygen and nitrogen.

In the same way that having a heteroatom present requires adjustment of the Coulomb integral along the diagonal, one might anticipate that the off-diagonal elements, which represent the resonance integrals, might also require adjustment. Thus, for a heteroatom having a shorter covalent radius, overlap with adjacent carbon p orbitals may be increased with a parallel increase in the absolute value of the resonance integral occurring in the off-diagonal element corresponding to interaction of the two atomic orbitals. Where such overlap is increased, the off-diagonal element ϵ is taken as greater than unity; where overlap with the heteroorbital is decreased, a value of ϵ of less than unity is utilized. Table 4.1-1 includes typical values for oxygen and nitrogen.

TABLE 4.1-1

	δ	ϵ
Ether oxygen	2.0	0.8
Ketone oxygen	1.0	1.0
Amine nitrogen	1.5	0.8
Imine nitrogen	0.5	1.0

The detailed rationale for the use of these new diagonal and off-diagonal elements is seen by considering a secular determinant such as that given in Eq. 2.2-12—however, now for a molecule having carbon atom 1 replaced by some heteroatom Z. The Coulomb integral for this atom now becomes

$$\alpha_z = \int \chi_z \mathcal{H} \chi_z \, dv$$

rather than the usual value for carbon

$$\alpha = \int \chi_c \mathcal{H} \chi_c \, dv.$$

The resonance integral linking the heteroatomic to the adjacent carbon atom(s) is now

$$\beta_{zc} = \int \chi_z \mathcal{H} \chi_c \, dv$$

rather than the usual

$$\beta = \int \chi_{cr} \mathcal{H} \chi_{cs} \, dv.$$

Secular determinant 2.2-12 then becomes

$$
\begin{array}{c}
 \\
\chi_1 \\
 \\
\chi_2 \\
 \\
\chi_3
\end{array}
\begin{array}{|ccc|}
\chi_1 & \chi_2 & \chi_3 \\
(\alpha_z - E) & \beta_{cz} & 0 \\
 & & \\
\beta_{cz} & (\alpha - E) & \beta \\
 & & \\
0 & \beta & (\alpha - E)
\end{array} = 0 \qquad\qquad 4.1\text{-}1
$$

If we make the substitution $(\alpha_z - E) = (\alpha_z - \alpha) + (\alpha - E)$ and then divide each column through by the constant factor β, the top left diagonal element becomes $(\alpha_z - \alpha)/\beta + (\alpha - E)/\beta$, or $(\delta + X)$ if we remember the definition of X given on page 52 and additionally define $\delta = (\alpha_z - \alpha)/\beta$. The remaining diagonal elements become X as in the absence of further heteroatoms. Also we define $\epsilon = \beta_{zc}/\beta$. Secular equation 4.1-1 then becomes

$$
\begin{array}{c@{\quad}c@{\quad}c}
 & \chi_1 \quad\; \chi_2 \;\;\; \chi_3 & \\
\begin{array}{c}\chi_1 \\[10pt] \chi_2 \\[10pt] \chi_3\end{array} &
\left|\begin{array}{ccc}
X + \delta & \epsilon & 0 \\[8pt]
\epsilon & X & 1 \\[8pt]
0 & 1 & X
\end{array}\right| = 0 &
\end{array}
\qquad 4.1\text{-}2
$$

It can be seen from the definition of δ that this quantity is the incremental value of the heteroatom Coulomb integral, in units of β, over that of an ordinary carbon p orbital. Since β is a negative unit, the lower the energy and the more negative the heteroatomic orbital energy, the more positive will be the value of δ.

The secular determinant in 4.1-2 is, of course, a special case. In the general situation there may be several heteroatoms. In these cases, each diagonal element corresponding to a heteroatomic orbital will have its own value of δ inserted, and each off-diagonal element will have a value of ϵ proportional to the resonance integral between the atomic orbitals heading the particular row and column.

4.2 Inclusion of Overlap

Hitherto in simplification of secular determinants (e.g., p. 51) we have been setting all overlap integrals $S_{rs} = 0$. This "neglect of overlap" made in the Hückel approximation was only partially justified by indicating that the overlap integrals between adjacent π-bonded carbon orbitals are small, ranging from 0.25 to 0.29 or so. However, this neglect is not entirely justified.[3]

If we are to include overlap, the simplest situation is encountered if it is assumed that all overlap integrals are equal. However, almost as manageable is the assumption[4] that the resonance integrals and the overlap integrals are proportional.

$$
\epsilon_{rs} = H_{rs}/\beta = S_{rs}/S \qquad \text{and} \qquad H_{rs} - ES_{rs} = \epsilon_{rs}(\beta - ES) \qquad 4.2\text{-}1
$$

where ϵ is a proportionality constant, β is the exchange integral for a standard bond length, and S is the corresponding standard overlap integral. The standard may be taken as that of benzene.

We may now use the definition of the off-diagonal elements $H_{rs} - ES_{rs}$, as given in Eq. 4.2-1, to substitute in a secular determinant such as 2.2-9. For the diagonal elements $(H_{rr} - ES_{rr})$ we substitute $(\alpha - E)$, as was done on page 51 in the case of neglect of overlap. These substitutions afford the secular determinant in 4.2-2* where all nonzero off-diagonal elements contain the common factor $(\beta - ES)$. Each column may be divided by this common factor without changing the equality to zero of the determinant. This affords a secular determinant of form similar to that of 2.2-13 where overlap was neglected. The new secular determinant is given in Eq. 4.2-3.

$$
\begin{array}{c|ccc}
 & \chi_1 & \chi_2 & \chi_3 \\
\hline
\chi_1 & (\alpha - E) & \epsilon_{12}(\beta - ES) & \epsilon_{13}(\beta - ES) \\
\chi_2 & \epsilon_{21}(\beta - ES) & (\alpha - E) & \epsilon_{23}(\beta - ES) \\
\chi_3 & \epsilon_{31}(\beta - ES) & \epsilon_{32}(\beta - ES) & (\alpha - E)
\end{array} = 0 \qquad 4.2\text{-}2
$$

$$
\begin{array}{c|ccc}
 & \chi_1 & \chi_2 & \chi_3 \\
\hline
\chi_1 & (\alpha - E)/(\beta - ES) & \epsilon_{12} & \epsilon_{13} \\
\chi_2 & \epsilon_{21} & (\alpha - E)/(\beta - ES) & \epsilon_{23} \\
\chi_3 & \epsilon_{31} & \epsilon_{32} & (\alpha - E)/(\beta - ES)
\end{array} = 0
$$

$$4.2\text{-}3$$

If we now let $X = (\alpha - E)/(\beta - ES)$ and substitute for the diagonal elements, we obtain a secular determinant of the same form as when overlap is neglected. Where the overlap and resonance integrals are taken equal to the standard S and β, respectively, then the off-diagonal elements become the usual 1's:

$$
\begin{vmatrix}
X & \epsilon_{12} & \epsilon_{13} \\
\epsilon_{21} & X & \epsilon_{23} \\
\epsilon_{31} & \epsilon_{32} & X
\end{vmatrix}
\quad \text{or} \quad
\begin{vmatrix}
X & 1 & 1 \\
1 & X & 1 \\
1 & 1 & X
\end{vmatrix} = 0 \qquad 4.2\text{-}4
$$

(for the special
case of
cyclopropenyl)

* It is seen that if atoms r and s do not overlap at all, ϵ_{rs} becomes zero.

While solution of the secular equations with retention of overlap thus leads to secular determinants of the usual form, the significance of the eigenvalues of X obtained no longer is the same as in the case of neglect of overlap as can be seen from the new definition of X above. Solving this definition of X for E, we find we may rearrange* the result to give a more meaningful form:

$$E = \frac{\alpha - X\beta}{1 - XS} = \frac{(\alpha - \alpha XS) + (\alpha XS - X\beta)}{1 - XS} = \alpha + \frac{X}{1 - XS}|\beta - \alpha S|$$

$$4.2\text{-}5$$

or

$$E = \alpha + X'|\gamma| \qquad\qquad 4.2\text{-}6$$

where

$$X' = \frac{X}{1 - XS} \qquad\qquad 4.2\text{-}7$$

and

$$\gamma = (\beta - \alpha S) \qquad\qquad 4.2\text{-}8$$

We note that Eq. 4.2-6 giving the energy when overlap is included has the same form as

$$E = \alpha + X|\beta| \qquad\qquad 4.2\text{-}9$$
$$(2.2\text{-}14\text{b})$$

which was derived on page 52 with neglect of overlap and which may be seen to result from 4.2-5 when S is set equal to zero.† In Eq. 4.2-6, with inclusion of overlap, we find that the energy is still expressed relative to α but the units of energy are now $|\beta - \alpha S|$ (i.e., $|\gamma|$). The eigenvalues of X as derived from the Hückel secular determinant are not linear with energy (E) and are not directly useful. However, by use of 4.2-7 we can use our Hückel eigenvalues (i.e., the X's) to give us the X''s which are directly linear with energy (Eq. 4.2-6).

Since the overlap integral between two adjacent p orbitals of a π system is relatively constant ($\sim$0.25), we may investigate how our new energy X'

* We note that both β and $(\beta - \alpha S)$ are negative numbers. Since it is convenient to work with positive units of energy, we take the absolute value of each to give energy units having the values $|\beta| = -\beta$ and $|\beta - \alpha S| = -(\beta - \alpha S)$.

† There is an interesting point here. When we make the approximation of setting $S = 0$ in Eq. 4.2-5, we cannot expect that the original value of β used will still give a satisfactory approximation to the energy E. Thus β of Eq. 4.2-5 and the β of Eq. 4.2-9 represent different quantities; the latter is no more than an empirical parameter assigned a value to correct for the approximations leading to 4.2-9. The former is a theoretical quantity, the resonance integral.

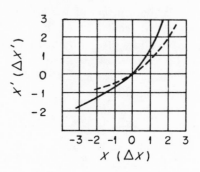

Fig. 4.2-A.   Plots of X' vs. X (solid curve) and $\Delta X'$ vs. ΔX (dashed curve).

varies with X. In addition and more pertinent is the way the calculated delocalization energy per electron compares in the two approaches. Plots of X' versus X and $\Delta X'$ versus ΔX are shown in Fig. 4.2-A. Here $\Delta X = (X + 1)$ and $\Delta X' = (X + 0.8)$ are the delocalization energies; the values 1 and 0.8 derive from our subtracting the values of X and X', respectively, for the bonding MO of ethylene as is necessary to obtain the *DE*s. We note that in the region of interest, namely, from X equal to ~-3 to about zero, both plots are only roughly linear. The slope of these plots determines what ratio of γ to β will successfully give the same π energies or one-electron delocalization energy for any given molecular orbital. Clearly, with the nonlinear plot, the ratio of γ to β optimum for one value of X will not be optimum for the other energy levels. However, often single molecular orbital energies are of less interest than the total π energy and the delocalization energy. Inspection of Table 4.2-1 shows that for the benzenoid compounds there is a good constancy of the γ-to-β ratio with a value of ~2. Accordingly, the commonly used values of $\beta = 18$ kcal/mole and $\gamma = 36$ kcal/mole not only reproduce the empirical resonance energy of benzene (36 kcal/mole) but also assure that calculation with or without overlap will give about the same delocalization energies for other benzenoid aromatics. For nonbenzenoid compounds, calculation using these values of β and γ will give only approximately the same results without as with overlap.

There is another consequence of inclusion of overlap; this might be termed "overlap destabilization." It can be seen that when overlap is included the bonding MOs are compressed toward zero (i.e., being nonbonding) and the antibonding MOs are expanded. Thus, looking at the expression for X' in Eq. 4.2-7, we see that for a negative eigenvalue of X, the denominator is greater than unity and X' becomes less negative than X. For a positive eigenvalue of X, the denominator becomes less than unity and X' becomes more positive than X. The net result is that where there is pairing of MOs (i.e., where a bonding and an antibonding pair of

TABLE 4-2.1

DELOCALIZATION ENERGIES WITH AND WITHOUT NEGLECT OF OVERLAP
FOR SELECTED COMPOUNDS

Compound	DE neglecting overlap	DE with overlap	Ratio of DEs (β/γ)
Butadiene	0.4721 β	0.1747 γ	2.702
Hexatriene	0.9879 β	0.3868 γ	2.554
3-Vinylhexatriene	1.466 β	0.5609 γ	2.578
Fulvene	1.466 β	0.6376 γ	2.299
Pentalene	2.456 β	1.087 γ	2.259
Heptalene	3.618 β	1.465 γ	2.471
Azulene	3.637 β	1.597 γ	2.278
Heptafulvene	4.005 β	1.496 γ	2.676
Fulvalene	2.799 β	1.204 γ	2.315
Styrene	2.424 β	1.211 γ	2.001
Stilbene	4.878 β	2.445 γ	1.994
1,1-Diphenylethylene	3.814 β	2.402 γ	2.004
Triphenylethylene	7.290 β	3.656 γ	1.994
Tetraphenylethylene	9.719 β	4.879 γ	1.992
Naphthalene	3.684 β	1.860 γ	1.981
Benzene	2.000 β	1.000 γ	2.000

MOs have opposite and equal values), the antibonding MO is more antibonding than the bonding MO is bonding. If both MOs are doubly occupied, there is net destabilization. This contrasts with the situation with neglect of overlap, where the stabilization due to two electrons in a bonding MO would exactly cancel the destabilization of two electrons in a paired antibonding MO.

For example, in the dianion formed from addition of two electrons to ethylene, neglect of overlap would give us zero delocalization energy whereas inclusion of overlap leads to net destabilization.

This result has been used to explain the reluctance of such systems to accept extra electrons.

4.3 Treatment of Hybrid and Unusually Oriented Orbitals

Hitherto we have utilized as a basis set of atomic orbitals only $2p$ orbitals oriented parallel-wise. There are a number of molecular systems of interest where the basis set includes $2s$ orbitals, sp hybrid orbitals, or $2p$ orbitals with other than the common parallel orientation. We wish now to deal with such situations.

4.3a   *The Form of Hybrid and Unusually Oriented Orbitals*

A particularly useful means of defining the direction of a vector in an x, y, z coordinate system is by its "direction cosines." These are the cosines α, β, and γ of the angles made by the vector with the x, y, and z axes, respectively (cf. Fig. 4.3-A).* Referring to this figure, we note that the coordinates of point P are given by

$$x = r\alpha, \qquad y = r\beta, \qquad \text{and} \qquad z = r\gamma \qquad\qquad 4.3\text{-}1$$

where r is the distance of point P from the origin. Equations 4.3-1 allow us now to rewrite the Slater atomic orbitals (Eq. 1.1-1) in more convenient form, a form useful in writing a general expression for hybrid and unusual orbitals. Thus the Slater atomic orbitals become

$$\chi_{2s} = \frac{k^{5/2}}{\sqrt{3\pi}} re^{-kr} \qquad\qquad 4.3\text{-}2$$

$$\chi_{2px} = \frac{k^{5/2}}{\sqrt{\pi}} \alpha re^{-kr} \qquad\qquad 4.3\text{-}3$$

$$\chi_{2py} = \frac{k^{5/2}}{\sqrt{\pi}} \beta re^{-kr} \qquad\qquad 4.3\text{-}4$$

$$\chi_{2pz} = \frac{k^{5/2}}{\sqrt{\pi}} \gamma re^{-kr} \qquad\qquad 4.3\text{-}5$$

Now a hybrid orbital will be an admixture of these four atomic orbitals

$$\phi = c_s\chi_s + c_{px}\chi_{px} + c_{py}\chi_{py} + c_{pz}\chi_{pz} \qquad\qquad 4.3\text{-}6$$

which on substitution of the Slater orbitals given in 4.3-2 to 4.3-5 becomes

$$\phi = \frac{k^{5/2}}{\sqrt{\pi}} re^{-kr} \left[\frac{c_s}{\sqrt{3}} + \alpha c_{px} + \beta c_{py} + \gamma c_{pz} \right] \qquad\qquad 4.3\text{-}7$$

Now we would like to know what assortment of constants ($c_s, c_{px}, c_{py}, c_{pz}$) will maximize the orbital in a given direction α, β, γ. Toward this end we may recognize that the orbital ϕ is a function of the direction cosines α, β, and γ and then extremize ϕ with respect to these variables. Since γ may be expressed in terms of α and β, i.e., $\gamma = (1 - \alpha^2 - \beta^2)^{1/2}$ Eq. 4.3-6 is

---

* We note that the three direction cosines are not mutually independent but are related by $\alpha^2 + \beta^2 + \gamma^2 = 1$. However, it does require all three to uniquely define the direction of vector OP. With only two cosines known, (e.g., α and β) two vectors symmetrically disposed about a plane (here XY) are possible.

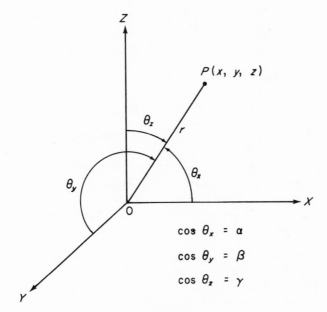

$$\cos \theta_x = \alpha$$
$$\cos \theta_y = \beta$$
$$\cos \theta_z = \gamma$$

FIG. 4.3-A

expressed as

$$\phi = \frac{k^{5/2}}{\sqrt{\pi}} \, re^{-kr} \left[\frac{c_s}{\sqrt{3}} + \alpha c_{px} + \beta c_{py} + c_{pz}(1 - \alpha^2 - \beta^2)^{1/2} \right] \qquad 4.3\text{-}8$$

Taking the partial derivatives with respect to α and β, setting these equal to zero for an extreme, and designating the values of the cosines giving the maximum as $\boldsymbol{\alpha}$, $\boldsymbol{\beta}$, and $\boldsymbol{\gamma}$, we obtain

$$\left[\frac{\partial \phi}{\partial \alpha} \right]_\beta = \frac{k^{5/2}}{\sqrt{\pi}} \, re^{-kr} \left[c_{px} - \frac{\boldsymbol{\alpha} c_{pz}}{(1 - \boldsymbol{\alpha}^2 - \boldsymbol{\beta}^2)^{1/2}} \right] = 0 \qquad 4.3\text{-}9$$

$$\left[\frac{\partial \phi}{\partial \beta} \right]_\alpha = \frac{k^{5/2}}{\sqrt{\pi}} \, re^{-kr} \left[c_{py} - \frac{\boldsymbol{\beta} c_{pz}}{(1 - \boldsymbol{\alpha}^2 - \boldsymbol{\beta}^2)^{1/2}} \right] = 0 \qquad 4.3\text{-}10$$

Using the relation $\alpha^2 + \beta^2 + \gamma^2 = 1$, we can solve these equations for the desired constants:

$$c_{px} = \boldsymbol{\alpha}(c_{pz}/\boldsymbol{\gamma}) \qquad 4.3\text{-}11$$

$$c_{py} = \boldsymbol{\beta}(c_{pz}/\boldsymbol{\gamma}) \qquad 4.3\text{-}12$$

$$c_{pz} = \boldsymbol{\gamma}(c_{pz}/\boldsymbol{\gamma}) \qquad 4.3\text{-}13$$

Thus the coefficients weighting the three different $2p$ orbitals are proportional to the direction cosines α, β, and γ defining the vector along which the orbital is maximized.* If we let the quantity (c_{pz}/γ) be represented by the symbol N_p, we may substitute the coefficients obtained in Eqs. 4.3-11 through 4.3-13 into 4.3-6 to write the form of a hybrid orbital aimed in the (α, β, γ) direction:

$$\phi = c_s\chi_s + N_p(\alpha\chi_{px} + \beta\chi_{py} + \gamma\chi_{pz}) \qquad 4.3\text{-}14$$

Thus we have here a general prescription for writing a hybrid orbital aimed in any given direction. It can be shown that for proper normalization $c_s^2 + N_p^2$ must equal unity. The s character of the hybrid is determined by the relative weightings of c_s and N_p. Specifically, if $n = c_s^2$ and $m = N_p^2$, then Eq. 4.3-14 gives an s^np^m hybrid.

A summary of the prescription for writing a hybrid orbital is then as follows:

1.   Write the orbital of the form given in Eq. 4.3-14 with the direction cosines α, β, and γ being selected to orient the orbital as desired.

2. c_s and N_p are chosen to give the wanted hybridization. The ratio of c_s^2 to N_p^2 affords the relative s to p character. For an s^np^m hybrid, take $c_s = \sqrt{n}$ and $N_p = \sqrt{m}$.

3.   Normalize the orbital. This is done by dividing the entire orbital by $(c_s^2 + N_p^2)^{1/2}$.

In order to better understand the reasons for the choice of c_s and N_p it is instructive to determine the energy of a hybrid orbital of the form given in Eq. 4.3-14. We do this by using the integrated form of the Schrödinger equation†

$$E = \int \phi\mathcal{3C}\phi \, dv = \int [c_s\chi_s + N_p(\alpha\chi_x + \beta\chi_y + \gamma\chi_z)]\mathcal{3C}[c_s\chi_s$$

$$+ N_p(\alpha\chi_x + \beta\chi_y + \gamma\chi_z)] \, dv. \quad 4.3\text{-}15$$

In the expansion of Eq. 4.3-15 we obtain integrals of the type $\int\chi_r\mathcal{3C}\chi_t \, dv$ where χ_r and χ_t are two different atomic orbitals of the same atom. These integrals are zero. That this is so may be seen by considering two examples. First consider the integral $\int\chi_s\mathcal{3C}\chi_z \, dv$. Implicit in the integral sign is the instruction to integrate from $x = -\infty$ to $+\infty$, $y = -\infty$ to $+\infty$, and

* When we substitute $\gamma = +(1 - \alpha^2 - \beta^2)^{1/2}$ we obtain a maximum. With the alternative of γ being taken negatively we find a minimum for ϕ. In the latter case c_{px}, c_{py}, and c_{pz} have their signs reversed and the orbital obtained is aimed oppositely with a negative lobe oriented in the α, β, γ direction.

† For simplicity α, β, and γ will now be used to represent the orbital orientation and χ_x, χ_y, and χ_z will be utilized for p orbitals.

$z = -\infty$ to $+\infty$. In this triple integration, there is no reason that one of the integrations cannot be split into two parts, an integration from (e.g.) $z = -\infty$ to 0 plus an integration from $z = 0$ to $+\infty$; it will be seen that these two parts are equal in magnitude but opposite in sign and hence self-canceling. Thus in the case of $\int \chi_s \mathcal{K} \chi_z \, dv$ for every point P^+ in the upper hemisphere (cf. Fig. 4.3-B) there is a point P^- below the xy plane where the sign and magnitude of χ_s is the same. But at P^- and P^+ χ_z and hence $\mathcal{K}\chi_z$ will have equal magnitudes but opposite signs. This derives from p orbitals having equivalent lobes of opposite sign. As a consequence, the value of $\chi_s \mathcal{K} \chi_z$ at P^- will be equal but opposite in sign to the value at P^+. The integration from $z = -\infty$ to 0 must afford a value equal but opposite in sign from the integration from 0 to $+\infty$; and the total integration from $z = -\infty$ to $+\infty$ will be zero. With similar reasoning we conclude that the integrals

$$\int \chi_s \mathcal{K} \chi_x \, dv, \quad \int \chi_s \mathcal{K} \chi_y \, dv, \quad \int \chi_x \mathcal{K} \chi_y \, dv, \quad \int \chi_x \mathcal{K} \chi_z \, dv \quad \text{and} \quad \int \chi_y \mathcal{K} \chi_z \, dv$$

are composed of AOs of different symmetry and are zero.

Returning to Eq. 4.3-15 and expanding, we may now discard cross-product-type integrals and are left with

$$E = c_s^2 \int \chi_s \mathcal{K} \chi_s \, dv + N_p^2 \left[\alpha^2 \int \chi_x \mathcal{K} \chi_x \, dv + \beta^2 \int \chi_y \mathcal{K} \chi_y \, dv + \gamma^2 \int \chi_z \mathcal{K} \chi_z \, dv \right]$$

$$4.3\text{-}16$$

The Coulomb integrals may be abbreviated as H_{ss} and H_{pp} and represent the energy of an electron in an isolated s or p orbital. Clearly the three Coulomb integrals involving χ_x, χ_y, and χ_z are equal since they differ only

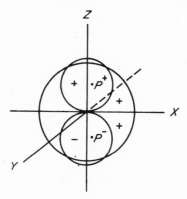

Fɪɢ. 4.3-B

in spatial orientation. Equation 4.3-16 then simplifies as

$$E = c_s^2 H_{ss} + N_p^2 [\alpha^2 + \beta^2 + \gamma^2] H_{pp} = c_s^2 H_{ss} + N_p^2 H_{pp} \qquad 4.3\text{-}17$$

We see now that the energy of the hybrid orbital has an s component and a p contribution and that these are weighted as c_s^2 and N_p^2. Herein lies the justification of our using the ratio of these squares as giving the $(s^n p^m)$ hybridization.

An exactly parallel expansion of the integral $\int \phi^2 \, dv = 1$ leads to the normalization requirement that $c_s^2 + N_p^2 = 1$.

Having discussed the basis for the rules given above for writing hybrid orbitals aimed at odd angles, we might illustrate use of this prescription. As a first example, suppose we wish to set up an sp^2 hybrid orbital centered at the origin and oriented in the $+Z$ direction. Since the orbital is perpendicular to the X and Y axes, α and β drop out of Eq. 4.3-14 (i.e., $\cos 90° = 0$). As a second step we note that the ratio of c_s to N_p must be $1:\sqrt{2}$. This would give us the unnormalized orbital $\phi' = \chi_s + \sqrt{2}\chi_z$ where c_s and N_p, before normalization, are 1 and $\sqrt{2}$, respectively. The third step of normalization requires division by the square root of the sum of squares of unnormalized c_s and N_p, that is, division by $\sqrt{3}$. Thus normalized $c_s = 1/\sqrt{3}$ and normalized $N_p = \sqrt{2}/\sqrt{3}$; and the normalized sp^2 hybrid orbital aimed in the positive z direction is $\phi = (1/\sqrt{3})\chi_s + (\sqrt{2}/\sqrt{3})\chi_z$. If one considers the spatial superposition of the two contributing atomic orbitals as in Fig. 4.3-C, one can see that the popular conception of an sp^2 orbital results (cf. Fig. 4.3-D). In the upper hemisphere in Fig. 4.3-C the positive lobe of the p_z atomic orbital adds to the positive value of the s orbital while in the lower hemisphere the negative value of the p_z orbital cancels the still positive value of the s orbital; the result is a large positive upper lobe and a small negative lower lobe.

Let us try the reverse approach in illustrating the writing of hybrid and oddly oriented orbitals. We can ask ourselves what sort of orbitals are given in the following:

$$\phi_1 = (1/\sqrt{6})\chi_s + (1/\sqrt{3})\chi_x + (1/\sqrt{2})\chi_z \qquad 4.3\text{-}18a$$

$$\phi_2 = (1/\sqrt{6})\chi_s + (1/\sqrt{3})\chi_x - (1\sqrt{2})\chi_z \qquad 4.3\text{-}18b$$

First of all, we can determine the sp hybridization from the ratio of the sum of squares of the p-orbital coefficients relative to the square of the s-orbital coefficient; this is equivalent to determining the energy contribution of the s- and p-type atomic orbitals. The ratio of s to p character is thus seen to be $\frac{1}{6} : (\frac{1}{3} + \frac{1}{2})$ or $\frac{1}{6} : \frac{5}{6}$ and both ϕ_1 and ϕ_2 are seen to be sp^5 hybrids. Second, we can concern ourselves with orientation in space. Knowing that c_s is $1/\sqrt{6}$ in each case and that $c_s^2 + N_p^2 = 1$, we recognize that

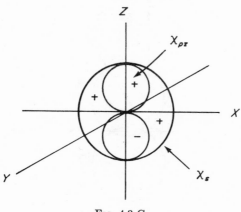

FIG. 4.3-C

when put into the form of Eq. 4.3-14 our orbitals ϕ_1 and ϕ_2 will have $N_p = \sqrt{5}/\sqrt{6}$. When we factor this out of the p-orbital portions of Eqs. 4.3-18a,b we obtain

$$\phi_1 = (1/\sqrt{6})\chi_s + (\sqrt{5}/\sqrt{6})[(\sqrt{2}/\sqrt{5})\chi_x + (\sqrt{3}/\sqrt{5})\chi_z] \qquad \text{4.3-19a}$$

$$\phi_2 = (1/\sqrt{6})\chi_s + (\sqrt{5}/\sqrt{6})[(\sqrt{2}/\sqrt{5})\chi_x - (\sqrt{3}/\sqrt{5})\chi_z] \qquad \text{4.3-19b}$$

In these equations we can find the direction cosines α, β, and γ as the respective coefficients of χ_x, χ_y, and χ_z. Since χ_y does not appear in these equations we can conclude that $\beta = 0$ and that the orbitals, being perpendicular to the Y axis, are in the XZ plane. For both ϕ_1 and ϕ_2 $\alpha = \cos \theta_x = \sqrt{2}/\sqrt{5}$; and therefore θ_x, the angle made by these orbitals with the X axis,

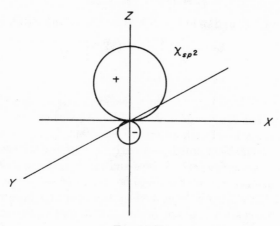

FIG. 4.3-D

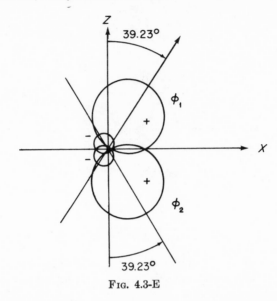

FIG. 4.3-E

is found to be 50.77°. For ϕ_1, $\gamma = \sqrt{3}/\sqrt{5}$ and $\theta_z = 39.23°$. For ϕ_2, $\gamma = -\sqrt{3}/\sqrt{5}$ and $\theta_z = 180° - 39.23°$. Thus we have a pair of sp^5 hybrid orbitals in the XZ plane and symmetrically disposed above and below the X axis as shown in Fig. 4.3-E.

There is an additional interesting aspect to the problem. Let us consider two new orbitals which are the normalized sum and difference of ϕ_1 and ϕ_2:

$$\phi_3 = (1/\sqrt{2})(\phi_1 + \phi_2) \qquad\qquad 4.3\text{-}20a$$

and

$$\phi_4 = (1/\sqrt{2})(\phi_1 - \phi_2) \qquad\qquad 4.3\text{-}20b$$

Substituting in for ϕ_1 and ϕ_2 in Eq. 4.3-20 we obtain the striking result that

$$\phi_3 = (1/\sqrt{3})\chi_s + (\sqrt{2}/\sqrt{3})\chi_x \qquad\qquad 4.3\text{-}21a$$

and

$$\phi_4 = \chi_z \qquad\qquad 4.3\text{-}21b$$

where ϕ_3 is seen to be an sp^2 hybrid directed positively along the X axis. It is seen that these are the orbitals used by one carbon atom of ethylene to form a σ and a π bond with a second such atom; ϕ_3 and ϕ_4 are depicted in Fig. 4.3-F. As has been noted previously, in forming molecular orbitals we arrive at the same final eigenvalues and eigenfunctions independent of the choice of the basis set of orbitals to be combined as long as a complete set is selected. Group orbitals formed as the sum and difference of two members of a basis set are as acceptable as the original two members. In the present instance we obtained ϕ_3 and ϕ_4 as the sum and difference of

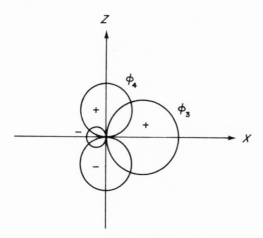

Fɪɢ. 4.3-F

ϕ_1 and ϕ_2. Actually, we could have started with ϕ_3 and ϕ_4, since as is easily shown the normalized sum and difference of these orbitals are ϕ_1 and ϕ_2. Thus for constructing the central σ and π bond of ethylene it makes no difference whether we use the familiar sp^2 orbital ϕ_3 to form the σ bond and then use the p_z orbital ϕ_4 (i.e., χ_z) to form the π bond or instead use two equivalent orbitals ϕ_1 and ϕ_2. The equivalent orbital representation of the double bond in ethylene is depicted in Fig. 4.3-G. Because these orbitals are aimed away from the interatomic axis and appear in the orbital picture drawn to interact to form a pair of orbitals splayed outward* the bonds thus formed have been termed "banana bonds." One might question whether the electron density directly between the two carbon atoms in this model would not be different from the ordinary $\sigma + \pi$ representation of a double bond, for Fig. 4.3-G seems to imply little such internuclear electron density, while the σ bond of the common model clearly provides such electron density. The difference is only apparent and derives from the diminu-

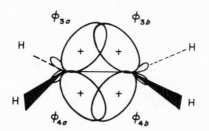

Fɪɢ. 4.3-G.   Basis set of sp^5 orbitals used in equivalent orbital double bond formulation.

* The similarity to the old spring model of a double bond is obvious.

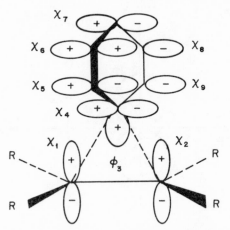

FIG. 4.3-H. Approximate model for a migrating phenyl group.

tive representation of atomic orbitals generally used by organic chemists. Such orbitals, as in Fig. 4.3-G, have the advantage of not cluttering up a molecular representation; however, a disadvantage results if one construes such drawings to imply the corresponding excessive electron localization. Thus, more liberally drawn orbital contours would be seen to provide the expected internuclear electron density. As a final point, it should be recognized that the basis set shown in Fig. 4.3-G will afford the same result as the sp^2 and p_x set only if all possible interaction is included in setting up the secular determinant. Thus the interaction between ϕ_{4a} and ϕ_{3b} and between ϕ_{3a} and ϕ_{4b} must be included although such interaction will be considerably smaller than between orbitals ϕ_{3a} and ϕ_{3b} and between ϕ_{4a} and ϕ_{4b}.

Having shown how to set up and identify miscellaneous hybrid and oddly oriented orbitals, it remains for us to show how these may be used in obtaining elements of secular determinants. As an example involving hybrid orbital and odd-angle orbital interaction let us select the model for the 1,2-phenyl migration depicted in Fig. 4.3-H.[5] The idealized model selected assumes cyclopropane distances between atoms bearing orbitals χ_1, χ_2, and χ_4. Also, all orbitals shown are taken as p except for ϕ_3 which is an sp^2 hybrid. The naiveté of these assumptions is justified only by lack of information about the precise hybridization and the recognition that the model will afford only qualitatively useful information.

By use of group theory as outlined in Chapter 3, we can set up the following symmetry orbitals:

$$A_1: \quad \chi_1 + \chi_2, \phi_3 \qquad A_2: \quad \chi_5 - \chi_9, \chi_6 - \chi_8$$

$$B_1: \quad \chi_1 - \chi_2, \chi_5 + \chi_9, \chi_6 + \chi_8, \chi_4, \chi_7$$

It is clear that to set up the A_1 and B_1 secular determinants, the interaction elements between orbitals ϕ_3 and χ_1 and χ_2 as well as between χ_4 and χ_1 and χ_2 will be needed. Since overlap integrals are easier to obtain than exchange integrals and since it is commonly assumed that these are proportional, the primary goal is to obtain the overlap integrals. The geometric relationships between χ_1 and χ_4 and between χ_1 and ϕ_3 are seen most readily by transposing these orbitals to the double coordinate systems of Figs. 4.3-I and 4.3-J. Our approach will be to write the analytical form for each of the hybrid and oddly oriented orbitals in Figs. 4.3-I and 4.3-J, and then to write down and expand the overlap integrals between pairs of these orbitals. The rationale behind this approach derives from the availability of tables of overlap integrals between common types of atomic orbitals oriented parallel-wise or perpendicularly and as a function of distance. These are the integrals which result from our expansion.

The orbitals whose linear combinations will be used are depicted in Fig. 4.3-K. In this figure as well as in 4.3-I and 4.3-J, we use a set of coordinates having two origins, a and b, 1.54 Å apart. At each origin there is a Z axis (Z_a and Z_b). The positive X directions are taken as facing the center so that the basis set will have the positive lobes of the orbitals directed toward each other. Each of the hybrid and oddly oriented orbitals of our bridged species (i.e., χ_1, χ_2, χ_4, and ϕ_3) may be written as a linear combination of the orbitals of Fig. 4.3-K; we use the geometry of Figs. 4.3-I and 4.3-J

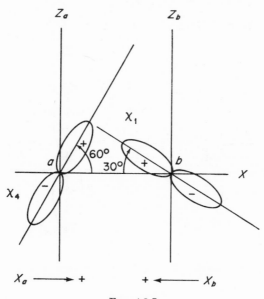

F ɪɢ. 4.3-I

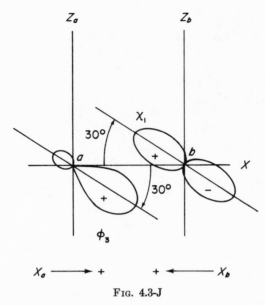

Fig. 4.3-J

together with our approach to obtain c_s, N_p, α, β, and γ.

$$\chi_1 = (\sqrt{3}/2)\chi_{xb} + \tfrac{1}{2}\chi_{zb} \qquad\qquad 4.3\text{-}22a$$

$$\chi_4 = \tfrac{1}{2}\chi_{xa} + (\sqrt{3}/2)\chi_{za} \qquad\qquad 4.3\text{-}22b$$

$$\phi_3 = (1/\sqrt{3})\chi_{sa} + (1/\sqrt{2})\chi_{xa} - (1/\sqrt{6})\chi_{za} \qquad\qquad 4.3\text{-}22c$$

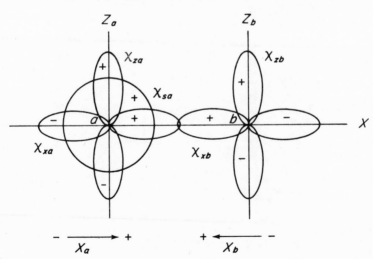

Fig. 4.3-K

Having the necessary orbitals formulated, let us proceed with determination of the required overlap integrals; for example,

$$S(\phi_3, \chi_1) = \int \phi_3 \chi_1 \, dv$$

$$= \int \left[(1/\sqrt{3})\chi_{sa} + (1/\sqrt{2})\chi_{xa} - (1/\sqrt{6})\chi_{za} \right]$$

$$\times \left[(\sqrt{3}/2)\chi_{xb} + \tfrac{1}{2}\chi_{zb} \right] dv$$

$$= \boxed{\tfrac{1}{2} \int \chi_{sa}\chi_{xb} \, dv} + \boxed{(\sqrt{3}/2\sqrt{2}) \int \chi_{xa}\chi_{xb} \, dv}$$

$$- (1/2\sqrt{2}) \int \chi_{za}\chi_{xb} \, dv + (1/2\sqrt{3}) \int \chi_{sa}\chi_{zb} \, dv$$

$$+ (1/2\sqrt{2}) \int \chi_{xa}\chi_{zb} \, dv - \boxed{(1/2\sqrt{6}) \int \chi_{za}\chi_{zb} \, dv}$$

$$= \tfrac{1}{2}S(\chi_{sa}, \chi_{xb}) + (\sqrt{3}/2\sqrt{2})S(\chi_{xa}, \chi_{xb})$$

$$- (1/2\sqrt{6})S(\chi_{za}, \chi_{zb}) \qquad\qquad \text{4.3-23a}$$

In the expansion of the overlap integral $S(\phi_3, X_1)$ all terms except those enclosed in boxes disappear due to differing symmetry of the two orbitals comprising the overlap integral. The reasoning is the same as on pages 134–135 in connection with resonance integrals involving orbitals at the same atom. Thus, for example

$$\int_{-\infty}^{0} \chi_{sa}\chi_{zb} \, dv = -\int_{0}^{\infty} \chi_{sa}\chi_{zb} \, dv; \qquad \text{and} \qquad \int \chi_{sa}\chi_{zb} \, dv = 0.$$

The remaining integrals of interest are obtained in the same way and are found to be

$$S(\chi_1, \chi_4) = (\sqrt{3}/4)S(\chi_{xa}, \chi_{xb}) + (\sqrt{3}/4)S(\chi_{za}, \chi_{zb}) \qquad \text{4.3-23b}$$

$$S(\chi_2, \chi_4) = S(\chi_1, \chi_4) \qquad\qquad\qquad\qquad\qquad\qquad \text{4.3-23c}$$

$$S(\phi_3, \chi_2) = S(\phi_3, \chi_1) \qquad\qquad\qquad\qquad\qquad\qquad \text{4.3-23d}$$

Thus, each of the specialized overlap integrals has been found as a linear combination of overlap integrals between regularly oriented orbitals, and the latter are conveniently available in tables, as has been noted. Table 4.3-1 contains values of special interest for organic calculations; these are

TABLE 4.3-1
SELECTED VALUES OF OVERLAP INTEGRALS[a]

Internuclear distance R (Å)	$S(\chi_{pz}, \chi_{pz})$	$S(\chi_{px}, \chi_{px})$	$S(\chi_{pz}, \chi_s)$	$S(\chi_s, \chi_s)$
0	1.00	−1.00	0	1.00
0.49	0.809	−0.483	0.386	0.890
0.98	0.468	0.159	0.509	0.637
1.24	0.318	0.303	0.463	0.491
1.30	0.287	0.319	0.444	0.456
1.37	0.258	0.328	0.425	0.423
1.43	0.221	0.332	0.402	0.390
1.50	0.207	0.332	0.380	0.360
1.56	0.184	0.327	0.357	0.330
1.63	0.164	0.319	0.334	0.302
1.95	0.089	0.250	0.226	0.188
2.28	0.046	0.171	0.141	0.111
2.60	0.023	0.107	0.083	0.062
2.93	0.011	0.063	0.046	0.034
3.26	0.005	0.035	0.025	0.018

[a] For $2s$ and $2p$ Slater orbitals on adjacent carbon atoms.

abstracted from the tables of Mulliken *et al.*[6] Thus by interpolating we obtain the orthodox integrals $S(\chi_{sa}, \chi_{xb})$, $S(\chi_{xa}, \chi_{xb})$, and $S(\chi_{za}, \chi_{xb})$ needed for Eqs. 4.3-23a–d, which then afford the desired overlap integrals $S(\chi_1, \chi_4)$, $S(\chi_2, \chi_4)$, $S(\chi_1, \phi_3)$, and $S(\chi_2, \phi_3)$. These overlap integrals may then be used to determine the off-diagonal elements of the secular determinant by assuming proportionality of these resonance integrals with the overlap integrals just found.* Although not part of the problem of determining the overlap integrals in unique situations, the results of the MO calculation on the bridged species in Fig. 4.3-H are interesting and deserve mention. Of the nine molecular orbitals found on solution of the secular equation, four are bonding and one is slightly antibonding besides the four strongly antibonding orbitals. Since the phenonium cation species corresponding to this structure has only eight delocalized electrons, the anti-bonding orbital is not utilized. However, in the corresponding free radical

* One might assume for qualitative purposes that the chief variation in secular determinant elements derives from differences in overlap.[6] However, a more detailed calculation would take into account the energy effects of s-orbital admixture. Diagonal elements corresponding to s-hybrids would be lower in energy by a difference which can be taken as the difference in the ionization potential of the $2s$ and $2p$ orbitals. The off-diagonal elements also would be adjusted by use of Mulliken's magic formula[7] which in the present case would give the off-diagonal element as $H_{ij} = \frac{1}{2}S_{ij}(H_{ii} + H_{jj})$. Other related versions[8] have been employed.

species and even more so in the carbanion analog, the antibonding orbital's contribution by virtue of occupation by one and two electrons, respectively, leads to destabilization. Nevertheless, a calculation similar to that just described, in which, however, an alkyl group migrates, indicates that the antibonding orbital is considerably higher in energy, with the radical and carbanion counterparts being less stable. This, of course, finds experimental support in organic chemistry.[5] We should note that the procedure followed in Eq. 4.3-23a may be generalized. We may assume orbitals of general form as given in Eq. 4.3-6, these orbitals being centered at points a and b. The overlap integral, determined as in the specific case, becomes

$$S_{ab} = c_{sa}c_{sb}S(sa, sb) + c_{xa}c_{xb}S(xa, xb) + c_{ya}c_{yb}S(ya, yb)$$

$$+ c_{za}c_{zb}S(za, zb) + c_{sa}c_{xb}S(sa, xb) + c_{xa}c_{sb}S(xa, sb) \qquad 4.3\text{-}24$$

We note that $S(ya, yb)$ and $S(za, zb)$ are equivalent; they represent the overlap between parallel p orbitals. We may call such an integral $S(p\pi, p\pi)$. Similarly, if we are dealing with carbon atoms, the last two overlap integrals are equal; we can designate these by $S(p\sigma, s)$. Here $p\pi$ or $p\sigma$ merely tells us whether the p orbital is perpendicular or coaxial with the interatomic axis (respectively). With the same descriptive notation the second integral can be briefly written $S(p\sigma, p\sigma)$. Then Eq. 4.3-24 is simplified to

$$S_{ab} = c_{sa}c_{sb}S(s, s) + c_{xa}c_{xb}S(p\sigma, p\sigma)$$

$$+ (c_{ya}c_{yb} + c_{za}c_{zb})S(p\pi, p\pi) + (c_{sa}c_{xb} + c_{xa}c_{sb})S(p\sigma, s) \qquad 4.3\text{-}25$$

Another convenient form is obtained by substituting (cf. pp. 133–134) $N_{pa}\alpha_a$ for c_{xa}, $N_{pa}\beta_a$ for c_{ya}, and so on. We obtain

$$S_{ab} = c_{sa}c_{sb}S(s, s) + N_{pa}N_{pb}[\alpha_a\alpha_b S(p\sigma, p\sigma) + \beta_a\beta_b S(p\pi, p\pi)$$

$$+ \gamma_a\gamma_b S(p\pi, p\pi)] + c_{sa}N_{pb}\alpha_b S(s, p\sigma) + c_{sb}N_{pa}\alpha_a S(s, p\sigma) \qquad 4.3\text{-}26$$

4.4   Properties of Alternant Hydrocarbons

Alternant hydrocarbons are defined as those π-system molecules where the carbon atoms may be separated into two sets, one which is starred and the other which is unstarred, and so that no two atoms of the same set are adjacent. Thus ethylene (I), butadiene (II), allyl (III), benzene (IV), and naphthalene (V) represent alternant systems while cyclopropenyl (VI), fulvene (VII), and azulene (VIII) are typical nonalternant molecules. It is seen in the nonalternant molecules that it is not possible to avoid having two starred or two unstarred atoms adjacent. Also, the common convention is to select the more numerous set of atoms for starring in the case of molecules having an uneven number of atoms; such molecules are termed odd alternant.

(I)   (II)   (III)   (IV)   (V)

Alternant

(VI)   (VII)   (VIII)

Nonalternant

There are a number of properties which are characteristic of alternant π systems. One is that alternant hydrocarbons (i.e., the neutral species) do not have any uneven electron density distribution and the π-electron density is unity at each center. The nonalternant molecules tend to have uneven electron density distributions; for example, both fulvene and azulene have large dipole moments.

A second property is that the MOs of the alternant hydrocarbons are symmetrically disposed about zero. A consequence of this is that the odd-alternant hydrocarbons then must have a nonbonding MO. This is so, since all of the MOs come in bonding–antibonding pairs, and one is left over; to be symmetrical the set then must have this remaining MO at zero.

Still another property of interest is that the LCAO-MO coefficients of alternant hydrocarbons are the same numerically for the bonding and antibonding pairs except that the sign is reversed at every unstarred atom. Alternatively we could reverse the sign of each starred orbital.

One further point deals with the odd-alternant hydrocarbons. This is that the free radical has its odd-electron density appearing at the starred atoms. The same is true of the positive charge of the carbonium and the negative charge of the corresponding carbanion.

The pairing theorems derive simply from a typical secular determinant of an alternant hydrocarbon. Thus, if we label the rows and columns corresponding to the starred atoms of the alternant system, we obtain in a typical (e.g., 4×4) system the secular determinant

$$
\begin{array}{c c c c c}
 & \chi_1{}^* & \chi_2 & \chi_3{}^* & \chi_4 \\
\chi_1{}^* & \begin{vmatrix} X \\ \\ a_{21} \\ \\ 0 \\ \\ a_{41} \end{vmatrix} & \begin{matrix} a_{12} \\ \\ X \\ \\ a_{32} \\ \\ 0 \end{matrix} & \begin{matrix} 0 \\ \\ a_{23} \\ \\ X \\ \\ a_{43} \end{matrix} & \begin{matrix} a_{14} \\ \\ 0 \\ \\ a_{34} \\ \\ X \end{matrix}
\end{array} = 0
$$

4.4-1

Here the a_{rs} elements merely represent the nonzero off-diagonal terms. The zeros present result from the fact that in an alternant hydrocarbon no two starred atoms will be adjacent and no two unstarred atoms will be adjacent, and there the orbitals which are starred correspond to starred atoms. If we now multiply the starred rows and columns by -1, we have in effect multiplied the determinant by -1 an even number of times, since each multiplication of a row or column of a determinant by a constant merely multiplies the value of the determinant by that constant. Thus we have not changed the equality to zero, and obtain

$$
\begin{array}{c c c c c}
 & -\chi_1{}^* & \chi_2 & -\chi_3{}^* & \chi_4 \\
-\chi_1{}^* & \begin{vmatrix} X \\ \\ -a_{21} \\ \\ 0 \\ \\ -a_{41} \end{vmatrix} & \begin{matrix} -a_{12} \\ \\ X \\ \\ -a_{32} \\ \\ 0 \end{matrix} & \begin{matrix} 0 \\ \\ -a_{23} \\ \\ X \\ \\ -a_{43} \end{matrix} & \begin{matrix} -a_{14} \\ \\ 0 \\ \\ -a_{34} \\ \\ X \end{matrix}
\end{array} = 0
$$

4.4-2

The eigenvalues of 4.4-2 are most readily seen from

$$
\begin{array}{c c c c c}
 & -\chi_1{}^* & \chi_2 & -\chi_3{}^* & \chi_4 \\
-\chi_1{}^* & \begin{vmatrix} -X \\ \\ a_{21} \\ \\ 0 \\ \\ a_{41} \end{vmatrix} & \begin{matrix} a_{12} \\ \\ -X \\ \\ a_{32} \\ \\ 0 \end{matrix} & \begin{matrix} 0 \\ \\ a_{23} \\ \\ -X \\ \\ a_{43} \end{matrix} & \begin{matrix} a_{14} \\ \\ 0 \\ \\ a_{34} \\ \\ -X \end{matrix}
\end{array} = 0
$$

4.4-3

which is just 4.4-2 negated; algebraically this is done by multiplying all columns and rows by the imaginary i which does not change the relative weightings of the basis functions.

In any case, it is seen that the eigenvalues of 4.4-3 are the negative of those of the original secular determinant 4.4-1, since the determinants of 4.4-1 and 4.4-2 differ only in having X's along the diagonal in the former and $-X$'s in the latter. Since 4.4-1 and 4.4-3 correspond to the same molecular problem, we see that for every eigenvalue X from 4.4-1, there is a corresponding eigenvalue $-X$ from 4.4-3. Thus the eigenvalues for the problem come in pairs equally displaced about the usual zero.

Turning to the eigenfunctions, we see that the eigenfunction from 4.4-1 corresponding to X and that from 4.4-3 corresponding to $-X$ will differ only in the signs of the coefficients of the starred basis orbitals. This can be seen from comparison of the orbitals heading the rows and columns in the two determinants. The only difference is in minus signs weighting the starred basis orbitals in 4.4-2.

4.5   The Dewar Nonbonding MO Method[9]

For odd-alternant hydrocarbons it has been noted that there is of necessity a nonbonding MO. Although one might not think that knowledge of this single MO of a system could be of use, a remarkable amount of information does result.

Refer to the secular equations of an odd-alternant hydrocarbon (e.g., benzyl; note Fig. 4.5-A and Eqs. 4.5-1a–4.5-1g):

$$c_1{}^*X + c_2{}^0 \qquad\qquad\qquad\qquad\qquad\qquad = 0 \quad \text{4.5-1a}$$

$$c_1{}^* \;+\; c_2{}^0X + c_3{}^* \qquad\qquad\qquad\qquad + c_7{}^* \;= 0 \quad \text{4.5-1b}$$

$$c_2{}^0 \;+\; c_3X^* + c_4{}^0 \qquad\qquad\qquad\qquad = 0 \quad \text{4.5-1c}$$

$$c_3{}^* \;+\; c_4{}^0X + c_5{}^* \qquad\qquad\qquad = 0 \quad \text{4.5-1d}$$

$$c_4{}^0 \;+\; c_5{}^*X + c_6{}^0 \qquad\qquad = 0 \quad \text{4.5-1e}$$

$$c_5{}^* \;+\; c_6{}^0X + c_7{}^* \;= 0 \quad \text{4.5-1f}$$

$$c_2{}^0 \qquad\qquad\qquad\quad\; + c_6{}^0 \;+ c_7{}^*X = 0 \quad \text{4.5-1g}$$

In these secular equations we use either * or ^{0} as the superscript to label the type of orbital weighted with the coefficient. This procedure reveals that the

Fig. 4.5-A.   The starred and unstarred positions of benzyl.

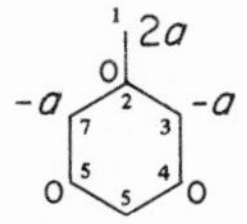

FIG. 4.5-B. Derivation of NBMO coefficients. $a = 1/\sqrt{7}$ by normalization.

LCAO-MO coefficients can be obtained essentially by inspection. Thus, we set all of the terms containing $X = 0$. The first equation then gives $c_2^0 = 0$. The second equation gives the sum of the coefficients bonded to the unstarred position 2, which equals zero; i.e., $c_1^* + c_3^* + c_7^* = 0$. Proceeding to the third secular equation and setting both X and c_2^0 equal to zero, we get the result that the unstarred coefficient $c_4^0 = 0$. Continuing in the same fashion we find in general that all of the unstarred coefficients for benzyl are zero and that in each case the sum of the starred coefficients surrounding any unstarred position adds up to zero. In fact, inspection of the general form of any set of secular equations such as 4.3-2 shows that when X in the rth secular equation is set equal to zero, there is left the sum of starred coefficients, and this sum generally equals zero. Similarly, it is general that the unstarred coefficients are zero.

With this generalization available, it is possible to obtain the nonbonding MO coefficients by inspection. Thus, we begin in the case of benzyl by assigning the value a to the para LCAO-MO coefficient. The coefficients at the meta positions are zero since these are unstarred positions. Now, the LCAO coefficient at carbon-5 is seen to be $-a$ since the sum of the coefficients surrounding C-4 (an unstarred position) must total zero. Similarly, the coefficient at C-7 is $-a$. Since two of the LCAO-MO coefficients surrounding C-2 are known and are $-a$, the coefficient for C-1 must be $+2a$. Finally, the coefficient at C-2 is zero since this is an unstarred carbon. Now, with the coefficients known relative to one another, and with the knowledge that the sums of their squares must equal unity by normalization, one can solve to obtain $a = 1/\sqrt{7}$.

NBMO coefficients can be obtained in general in the same manner (Fig. 4.5-B). A useful point to keep in mind is that it is usually simpler to begin by assigning the unknown value a to a starred position which is distant from the carbon bonded to only a single unstarred position. If we had begun by assigning a value to the coefficient at C-1 first, we could have obtained the same answer but with slightly more difficulty.

4.6 Nonbonding MOs in Möbius Systems

Dewar's treatment given in Section 4.5 was predicated on the assumption of a simple π system with no plus–minus overlaps in the basis set. An

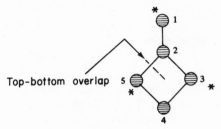

FIG. 4.6-A

exactly parallel treatment of an odd-alternant system having a sign discontinuity (i.e., a plus–minus overlap) in the basis set leads us to a result differing only slightly. For example, consider the species having a Möbius cyclobutadiene ring and one exocyclic atom (note Fig. 4.6-A).
The secular equations are

$$c_1^*X + c_2^0 \qquad\qquad\qquad\qquad = 0 \qquad\qquad 4.6\text{-}1$$

$$c_1^* \ + c_2^0X + c_3^* \qquad\qquad - c_5^* \ = 0 \qquad\qquad 4.6\text{-}2$$

$$c_2^0 \ + c_3^*X + c_4^0 \qquad\qquad = 0 \qquad\qquad 4.6\text{-}3$$

$$c_3^* \ + c_4^0X + c_5^* \ = 0 \qquad\qquad 4.6\text{-}4$$

$$- c_2^0 \qquad\qquad c_4^0 \ + c_5^*X \ = 0 \qquad\qquad 4.6\text{-}5$$

Realizing that we can star alternant atoms, we know that pairing of MOs occurs for such Möbius systems. In odd systems as presently considered there then must be a nonbonding MO. With X set equal to zero, we can see that the unstarred coefficients again must be zero as for the Hückel and acyclic situations. For example, in Eq. 4.6-1, with $X = 0$, c_2 becomes zero. However, Eq. 4.6-2 shows a slight difference from the normal Dewar treatment. Here, with $X = 0$, $c_1 + c_3 - c_5 = 0$. Thus the rule becomes that the coefficients surrounding any unstarred position are added; however, in this addition they are taken with a negative sign if they overlap with that unstarred basis orbital in a plus–minus fashion. With ordinary overlap the starred coefficients are added as usual with a positive sign. The sum must be zero as in the normal Hückel cases. We can then generalize the Dewar rule as

$$\sum_r \epsilon_{rs} c_r^* = 0 \qquad\qquad 4.6\text{-}6$$

where s is the unstarred atom, the summation is over all overlapping atoms r, and ϵ_{rs} is $+1$ for plus–plus or minus–minus overlap between basis orbitals r and s while it is -1 for plus–minus overlap.

4.7 Uses of the NBMO Coefficients

A number of different practical results can be derived with knowledge of
the NBMO coefficients.[9] One allows us to obtain the electron distribution
in the odd-alternant carbanion or carbonium ion. Thus, in the benzyl
radical, having seven π electrons, we have a unit π-electron density at all
atoms of the π system as is characteristic of all the uncharged alternant
hydrocarbons. In looking at Fig. 4.7-A, we see that the carbanion differs
only in the addition of one extra nonbonding MO electron, and we know
the distribution of this electron from the LCAO-MO coefficients we have
derived. The squares of these coefficients give the electron densities which
are then seen to be $\frac{4}{7}$ at the benzylic carbon (C-1) and $\frac{1}{7}$ at each of the
ortho and para carbons. These values then correspond to the distribution
of formal negative charge in this species. In the case of the corresponding
cation, we have the same configuration as in the free radical except that
the single nonbonding electron is removed. Since the distribution of this
nonbonding electron is known, we then know the distribution of formal
positive charge in the cation. It is again $\frac{4}{7}$ benzylic and $\frac{1}{7}$ ortho and para.
The results obtained are identical to those resulting from complete solu-
tion of the Hückel secular determinant.

A number of other uses of the nonbonding MO coefficients have been
demonstrated, especially by Dewar.[9] One is estimation of the energy gained
by juxtaposition of two odd-alternant hydrocarbon fragments to generate
an even-alternant system of interest. As an approximation, we assume that
the energy resulting from fusion of the two fragments derives from mixing
of the two nonbonding MOs, one from each fragment and with the two
electrons then populating the lower energy of the two MOs. In effect then
we need to mix together the two NBMOs ψ_A and ψ_B in a 2×2 secular

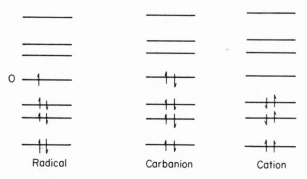

FIG. 4.7-A. Configurations of the benzyl radical, carbanion, and carbonium ion.

determinant. It can be readily seen that the determinant has off-diagonal elements as in the following equation:

$$
\begin{array}{c}
\quad\quad \psi_A \quad\quad\quad \psi_B \\
\begin{array}{c} \psi_A \\ \\ \psi_B \end{array}
\left|
\begin{array}{cc}
X & (c_{rA}c_{sB}) \\
\\
(c_{rA}c_{sB}) & X
\end{array}
\right| = 0
\end{array}
\qquad 4.7\text{-}1
$$

and with X's along the diagonal. Actually with normalized MOs, here ψ_A and ψ_B, the number of squared terms obtained along the diagonal is always unity. The product $c_{rA}c_{sB}$ is based on the assumption that fragments A and B overlap only at one place, where atom r of A overlaps with atom s of B. If there are more sites of overlap, then we have a sum of such terms off the diagonal. In any case, the off-diagonal term is arrived at in the usual way in which we consider the number of adjacent overlaps.

Solving the 2×2 thus obtained gives us

$$
X = \pm[c_{rA}c_{sB} + c_{tA}c_{uB} + \cdots] \qquad 4.7\text{-}2
$$

Since the splitting of the two nonbonding MOs gives a bonding MO which is doubly occupied, the energy lowering due to this splitting is

$$
\Delta E = -2(c_{rA}c_{sB} + 2c_{tA}c_{uB} + \cdots) \qquad 4.7\text{-}3
$$

In words, this means that we can put two odd-alternant fragments together and approximate the stabilization energy by taking the product of the NBMO coefficients, one from each fragment, at each site of new bonding. The sum of products is then doubled and taken negatively. We note here that the absolute sign of LCAO-MO coefficients is arbitrary, so that after the summation we may end up with a positive or negative summation, but this has no significance.

One example which is of some interest has been noted by Dewar. This is the formation of naphthalene and azulene from the nonatetraenyl and the one-carbon fragments:

$$4.7\text{-}4$$

$$4.7\text{-}5$$

Here b derives from a single p-orbital fragment and is therefore 1 on normalization. The value of a is obtained in the usual way as $1/\sqrt{5}$. However, one does not need the actual values of a and b in order to see that there is energy lowering at two sites in azulene compared with three in the formation of naphthalene. The energy changes on formation of these two species are thus given in the equations above. The main difference results from the fact that all sites of bonding have the same contribution of $2ab$ in the case of naphthalene but for azulene one term less occurs due to bonding at an unstarred position. Even worse energetically would be an example where bonding led to terms with opposite sign as in the formation of cyclobutadiene from a methyl fragment and allyl:

$$\Delta E = 0 \qquad\qquad 4.7\text{-}6$$

Here we have one contribution of $+2ab$ and one negative contribution of $-2ab$. Although b is still unity, the value of a in this case is not the same as in the preceding example but is the usual $1/\sqrt{2}$ (i.e., the nonbonding MO coefficients of allyl).

4.8 The Mulliken–Wheland–Mann and Omega Techniques

Thus far in doing MO calculations we have assumed that all adjacent overlaps are equal and that all carbon atoms are equally electron attracting. Although this simplistic set of assumptions is incorporated in the secular determinant prior to solution, the results of such Hückel calculations are less naive. Thus, bond orders do not come out all equal except where demanded by symmetry. Similarly, in nonalternant hydrocarbons and also in charged species, one finds unequal electron densities at different molecular sites.

One attempt to take this information into account in setting up the initial secular determinant was proposed by Mulliken and Wheland and students.[4,10,11] Thus it is assumed that the resonance integral used, representing interaction between two atoms, should be linearly dependent on the bond order between the two atoms. Similarly, it is assumed that the Coulomb integral used should be scaled so that it is a function of the electron density at each atom. The relationships used are

$$\delta_r = \omega(1 - Q_r) \qquad\qquad 4.8\text{-}1$$

$$\beta_{rs} = \beta_0(S_{rs}/S_0) \qquad\qquad 4.8\text{-}2$$

where

$$S_{rs}/S_0 = [0.08(P_{rs}{}^{\pi} + P_{rs}{}^{\sigma}) + 0.115]/0.276 \qquad 4.8\text{-}3$$

Here Q_r is the total π-electron density at carbon r, β_0 is the standard resonance integral (e.g., for ethylene), β_{rs} is the adjusted resonance integral between atoms r and s, $P_{rs}{}^{\pi}$ is the π bond order between atoms r and s, while $P_{rs}{}^{\sigma}$ is generally taken as unity. Also, δ_r is the increment in the diagonal element of the secular determinant used earlier for an atom r of different electronegativity than carbon. It can be seen that both resonance integrals and diagonal secular determinant elements are adjusted. For a more electronegative atom than carbon, a positive δ was employed. Equation 4.8-1 indicates that δ is taken as a function of the π-electron density at the given carbon. If this electron density, Q_r, is less than the unit density found for an uncharged alternant hydrocarbon, then δ will be positive, since the proportionality factor ω is positive; a typical value of ω is 0.8. If the electron density is greater than unity at a carbon, the value of δ will be negative. However, one can determine the electron density only from the LCAO-MO coefficients and these are not available *prior to* setting up the secular determinant. Thus a reiterative method is needed. Similarly, Eqs. 4.8-2 and 4.8-3 allow one to obtain new off-diagonal elements (i.e., the β_{rs}'s) if one uses the bond orders derived from the preceding iteration. The first iteration is an ordinary Hückel solution of the MO problem. The LCAO-MO coefficients obtained are then used via Eqs. 4.8-1, 4.8-2, and 4.8-3 to obtain the elements of the secular determinant for a second iteration. This determinant, on solution, then gives new coefficients which are then used again. The process is continued until the eigenvalues and coefficients reach constant values.

While Mulliken, Wheland, and co-workers used both variations of the Coulomb integrals and resonance integrals, Streitwieser has shown that in many instances variation of only the diagonal elements provides a useful improvement. This has been termed the omega technique.[12]

Another approach is to use only the variation of off-diagonal elements with bond orders, and this has been used by the present author. One example of the utility of this last approach is in the application to the pentadienyl anion. Here a simple Hückel calculation suggests that the π-electron densities at carbons-2 and -4 are zero but 0.333 at each of atoms 1, 3, and 5. Reasons have been given[13] as to why the electron density seems experimentally to be more heavily localized at the center carbon of such systems in disagreement with this simple calculation. However, the Hückel approach is less naive in giving bond orders. The bond order between atoms 1 and 2 is calculated to be different than that between 2 and 3. Furthermore, the improved bond orders are seen to lead to the correct concentration of

TABLE 4.8-1

π-Electron Densities Calculated for the Pentadienyl Carbanion by Successive LCAO-MO Approximations with Changing Bond Orders[a]

Approximation	q_1	q_2	q_3	P_{12}	P_{23}
First	0.333	0.000	0.333	0.788	0.578
Second	0.317	0.000	0.365	0.802	0.564
Third	0.316	0.000	0.368	0.802	0.562

[a] Adopted from Zimmerman.[13]

charge at the central carbon. The results of successive approximations for the pentadienyl anion are given in Table 4.8-1.

4.9 Correlation Diagrams; Reaction Allowedness and Forbiddenness

One application of molecular orbital theory which is of particular use to the organic chemist is the prediction of the "allowedness" or "forbiddenness" of an organic reaction. One suitable definition is that allowed ground-state reactions are those in which there are no crossings of MOs during reaction except where the two MOs crossing have the same occupation of electrons. This means that the product configuration will have the electrons in the lowest possible arrangement. For a reactant with only bonding electrons (e.g., a neutral, nonradical hydrocarbon) this means that an allowed reaction will be one in which all of the bonding MOs remain bonding during reaction and all of the antibonding MOs remain antibonding. This situation is shown in Fig. 4.9-A.

Conversely, a forbidden reaction is one in which, as the reaction proceeds, the MOs cross in such a way that one ends up with a higher energy electron population of product MOs than one had in the reactant. For a

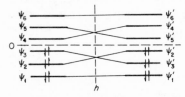

FIG. 4.9-A. A typical allowed reaction. All bonding MOs remain bonding. The abscissa is the reaction coordinate and the ordinate is the energy X.

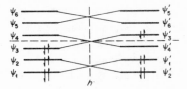

FIG. 4.9-B. A typical forbidden reaction. Bonding and antibonding MOs cross $X = 0$. The abscissa and ordinate are the same as in Fig. 4.9-A.

simple reactant with only bonding electrons this means that there will be a bonding MO (with two electrons) which becomes antibonding and an antibonding MO (unoccupied) which becomes bonding. This situation is illustrated in Fig. 4.9-B. Here we see that in the allowed reaction (Fig. 4.9-A), all the bonding MOs remain bonding and the product is obtained in its lowest configuration with all the bonding MOs occupied but no antibonding MOs containing electrons. In contrast, in the forbidden reaction (Fig. 4.9-B) ψ_3 of reactant gradually transforms itself into ψ_3' of product and ψ_3' is antibonding. Since ψ_3 is doubly occupied, assuming that the reaction is adiabatic (i.e., there is no change in electron occupation), ψ_3' will be doubly occupied as well and thus an (exceptionally high energy) excited state of product would result. While in such forbidden reactions one does not expect to get a doubly excited state as the product, one can expect that the molecular energy will rise sharply as the molecule proceeds along the reaction coordinate.

Now the question is how one determines which MOs of reactant become which MOs of product. One approach can be used in cases where there is molecular symmetry in the reactant which does not change along the reaction coordinate. It is important, though, to note that it is not sufficient for reactant and product merely to have the same symmetry but rather it is necessary for the symmetry to be maintained in between these extreme molecular geometries. In such cases, one is able to draw the correlation lines (i.e., as in Figs. 4.9-A and 4.9-B) by use of the noncrossing rule which states that two eigenfunctions (here two MOs) of the same symmetry will not cross. It is found that there is one unique correlation possible in each case which satisfies the criterion. This method is the one introduced by Woodward and Hoffmann.[14,15]

As an example, we can apply the method to the electrocyclic closure of allyl. The orbital drawings at the top of Fig. 4.9-C are for the basis set. We see that the disrotatory motion allows the allyl species to maintain a vertical plane of symmetry as twisting continues until finally the cyclopropyl species is generated. Conversely, conrotatory motion maintains a horizontal axis of symmetry throughout the process. The symmetries are designated as *A* or *S* representing antisymmetry or symmetry with respect to these elements (i.e., the plane or axis). We find only one way to connect reactant and product MOs in each case.

If we now consider a given occupation of electrons—two electrons for allyl cation giving cyclopropyl cation or four electrons for allyl anion giving cyclopropyl anion—we find that the disrotatory motion is favored for the cation since here there is no crossing of MO 1 with any unoccupied MO as does happen in the conrotatory twisting. For the anion, with four delocalized electrons in MOs 1 and 2, we see that conrotatory motion is preferred

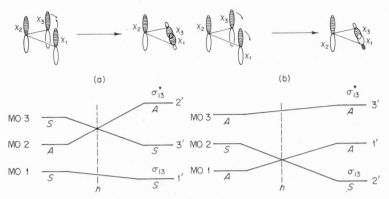

FIG. 4.9-C. Correlation diagrams for disrotatory and conrotatory electrocyclic closures of allyl species. (a) Disrotatory twisting of basis set; (b) conrotatory twisting of basis set.

since here MOs 1 and 2 are each doubly occupied; these cross in the conrotatory mode and we obtain a ground-state configuration. Alternatively, if we had disrotatory motion, the two electrons in MO 2 become strongly antibonding, since with this motion MO 2 becomes the antibonding σ orbital of product; also, MO 3 of reactant is unoccupied and transforms itself into the nonbonding MO of the carbanion. Thus, disrotatory motion leads to a cyclopropyl anion with two electrons in an antibonding σ orbital but none in the p orbital at atom 2. Hence, for the carbanion conrotatory motion is allowed and disrotatory motion is forbidden.

Another approach to drawing the correlation lines and following energy change along a reaction coordinate involves looking at the reacting species half-way along the reaction coordinate. This is the Möbius–Hückel method of Zimmerman.[16] For example, if we refer to Figs. 4.9-A and 4.9-B, we see that half-way between reactant and product there are degeneracies at the point along the reaction coordinate where MOs cross. This half-way point along the reaction coordinate is marked with an h. What we need then is a way to determine what the array of MOs looks like at half-reaction, since for every degeneracy there is a crossing of MOs and this would allow us to draw the correlation diagram.

Turning to the specific case of the allyl species closure, we note that at half-reaction the disrotatory transition state is composed of a Hückel-like array of basis orbitals, since nowhere do we have a plus–minus overlap between basis MOs (see Fig. 4.9-D). As noted earlier in connection with basic MO theory, the MOs deriving from mixing of the basis orbitals are independent of our choice of orientation of the basis orbitals; and here our definition of Hückel and Möbius is independent of just how we orient the

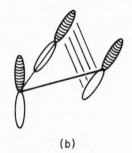

(a)   (b)

FIG. 4.9-D.   Disrotatory (Hückel) and conrotatory (Möbius) transition states for allyl closure. (a) Hückel transition state; (b) Möbius transition state.

basis orbital selected. If we invert one (e.g., p orbital) in the set of either system, we do not change the Hückel or Möbius character. The former always has zero or an even number of inversions in proceeding around the cyclic array and the latter has an odd number of plus–minus overlaps.

Presently, we know that a Möbius cyclopropenyl system has a bonding degeneracy and a single antibonding MO. This tells us that MOs 1 and 2 cross at the point h, as actually occurs in Fig. 4.9-C. In contrast, the Hückel cyclopropenyl system has a single bonding MO and an antibonding degeneracy. This tells us that for the Hückel reaction geometry (i.e., disrotatory), MOs 2 and 3 cross, as does happen.

Thus the Möbius–Hückel approach of Zimmerman[16] allows one to obtain correlation diagrams readily, and the method is not dependent on symmetry.

Problems

1.   Using symmetry, derive the correlation diagram for the disrotatory closure of the allyl cation to give the cyclopropyl cation. Now do the same for the conrotatory closure.

2.   Use the Hückel–Möbius method to do the two problems in Problem 1.

3.   Use the bond order–perturbation method to draw the correlation diagram for this same reaction. Have you obtained the same result from the three different approaches? In each case indicate if the reaction is forbidden or allowed.

4.   Use the LCAO-MO coefficients given for butadiene in Chapter 1 to determine if either disrotatory twisting or conrotatory twisting is forbidden. Do the other approaches give the same result?

5. Consider the transformation of "rectangular cyclobutadiene," in which the long bonds are 1–2 and 3–4 and the short bonds are 2–3 and 1–4, into its isomer in which the short bonds have been stretched into the long bonds and the long bonds have become the short bonds (i.e., the long bonds now are 2–3 and 1–4 and the short bonds are now 1–2 and 3–4). Obtain the correlation diagram and show that this is a forbidden transformation. [*Hint*: One approach is to use symmetry and another is to use the Möbius–Hückel approach to obtain the half-reaction MO array.] Generalize this; for example, what is the result with cyclooctatetraene? How about other $4N$ systems? If only π-electronic factors are controlling, what is the preferred geometry?

6. In view of the result of Problem 5, inspect a similar transformation of a $4N + 2$ system such as benzene, here a benzene where bonds 1–2, 3–4, and 5–6 are short and 2–3, 4–5, and 6–1 are long in one species while 1–2, 3–4, and 5–6 are long and 2–3, 4–5, and 6–1 are short in the other. Is this transformation forbidden or allowed? What is the preferred molecular geometry of (e.g.) benzene as a consequence.

7. Categorize the transition states for each of the following transformations as either Möbius or Hückel. Then determine the number of delocalized electrons involved in each of the transition states. Finally, draw the correlation diagram and decide if the reaction is forbidden or allowed.

(a) A disrotatory electrocyclic closure of butadiene to give cyclobutene.

(b) The conrotatory version of part (a).

(c) A 1,3-antarafacial hydrogen migration in propylene.

(d) A 1,3-suprafacial hydrogen migration in propylene.

(e) A process analogous to that in part (c), however, where a carbon with a p orbital is doing the 1,3-migrating and both lobes of the p orbital are overlapping, one lobe with the top of C-1 and the other lobe with the bottom of C-3.

8. Consider the 1,4-closure of benzene to Dewar benzene with top–top overlap between AOs 1 and 4 of benzene. Draw the correlation diagram using symmetry. Is the reaction forbidden or allowed? Now consider the problem from another approach. Dissect the two MOs which are antisymmetric with respect to a plane bisecting carbon atoms 1 and 4 and note that two of the six delocalized electrons populate one of these two; also note that this MO remains bonding throughout the reaction and does not affect the allowedness or forbiddenness. Now inspect the remaining four MOs and note that each of these can be written as composed of group orbitals of the form χ_1, $(\chi_2 + \chi_6)$, $(\chi_3 + \chi_5)$, χ_4 and that these four basis group orbitals form a cyclic array and must accommodate the remaining four delocalized electrons. What type of array is this, Hückel or Möbius? Is the reaction allowedness or forbiddenness predictable on this basis? Can you now generalize the Möbius–Hückel treatment to such bicyclic transition state reactions?

9. Using the Dewar method of nonbonding MO coefficients, predict the charge distribution on the methylene carbons of (a) the benzyl anion and (b) the α-naphthylmethyl anion (i.e., α-naphthyl-$CH_2^{\ominus}$:) in order to decide which is more basic.

10. Draw the correlation diagram for the thermal fission of cyclobutane into ethylene. Label the diagram explicitly and conclude if the reaction is forbidden or allowed.

References

1.   L. Pauling and G. W. Wheland, *J. Amer. Chem. Soc.* **57**, 286 (1935).
2.   A. Streitwieser, "Molecular Orbital Theory for Organic Chemists." Wiley, New York, 1961; a particularly broad survey is given here.
3.   R. S. Mulliken, C. A. Rieke, and W. G. Brown, *J. Amer. Chem. Soc.* **63**, 1770 (1941).
4.   G. Wheland, *J. Amer. Chem. Soc.* **64**, 900 (1942).
5.   H. E. Zimmerman and A. Zweig, *J. Amer. Chem. Soc.* **83**, 1196 (1961).
6.   R. S. Mulliken, C. A. Rieke, D. Orloff, and H. Orloff, *J. Chem. Phys.* **17**, 1248 (1949).
7.   R. S. Mulliken, *J. Phys. Chem.* **56**, 292 (1952); also, *J. Chim. Phys.* **46**, 497, 675 (1949).
8.   M. Wolfsberg and L. Helmholtz, *J. Chem. Phys.* **20**, 837 (1952).
9.   M. J. S. Dewar, "The Molecular Orbital Theory of Organic Chemistry." McGraw-Hill, New York, 1969, p. 212
10   N Muller and R. S. Mulliken, *J. Amer. Chem. Soc.* **80**, 3489 (1958).
11. N. Muller, L. W. Pickett, and R. S. Mulliken, *J. Amer. Chem. Soc.* **76**, 4770 (1954).
12. A. Streitwieser, Jr. and P. M. Nair, *Tetrahedron* **5**, 149 (1959).
13. H. E. Zimmerman, *Tetrahedron* **16**, 169 (1961).
14. (a) R. B. Woodward and R. Hoffmann, *Angew. Chem., Int. Ed. Engl.* **8**, 781 (1969); (b) R. B. Woodward and R. Hoffmann, *J. Amer. Chem. Soc.* **87**, 395 (1965); (c) R. B. Woodward and R. Hoffmann, *ibid.* **87**, 2511 (1965); (d) R. Hoffmann and R. B. Woodward, *ibid.* **87**, 2046 (1965); (e) R. Hoffmann and R. B. Woodward, *ibid.* **87**, 4389 (1965).
15. See also, H. C. Longuet-Higgins and W. W. Abrahamson, *J. Amer. Chem. Soc.* **87**, 2045 (1965).
16. (a) H. E. Zimmerman, *J. Amer. Chem. Soc.* **88**, 1564 (1966); *Accounts Chem. Res.* **4**, 272 (1971); **5**, 393 (1972).

Chapter 5

MORE ADVANCED METHODS; THREE-DIMENSIONAL TREATMENTS AND POLYELECTRON WAVEFUNCTIONS

Thus far we have used MO methods which apply to a truncated set of basis orbitals. Either the truncated set has been the π system of a planar molecule where then we neglect the σ system, or the set has been a group of orbitals assumed to be able to be dissected from the rest of the molecule. Also, hitherto we have included only one electron at a time in our energy minimization and have not included electron–electron interaction effects. This chapter deals with the polyelectron methods and the extension of Hückel theory to three dimensions.

5.1 Polyelectron Wavefunctions; Slater Determinants

Thus far we have been considering wavefunctions which are single molecular orbitals, each of which is considered as contributing separately and independently to the state of the molecule. However, a better wavefunction would consist of a product of single molecular orbitals and have a spin assignment as well as having electrons assigned to molecular orbitals in the product. This would then give the simultaneous contribution of all MOs. Such a product would be

$$\Phi = \Psi_1(1)\bar{\Psi}_1(2) \qquad \text{5.1-1}$$

for ethylene. Here the presence of the bar over the MO indicates that a β spin is assigned to that MO and the absence of a bar indicates an α spin for that MO. This product, in effect, says that while electron 1 is assigned to MO 1 with a requirement for α spin, electron 2 is assigned to MO 1 with imposition of β spin. The overall wavefunction then is the product of the two space-spin orbitals. It can be seen that the wavefunction squared, which gives the probability and electron distribution at any point in space, is then just the product of the individual MOs squared and thus just the product of the individual one-electron probabilities.

Each of the constituent orbitals is called a space-spin orbital, or often just a spin orbital, since it describes both the spatial and the spin properties of an electron assigned to it. What is meant is that the orbital is a product of the ordinary spatial MO of the type we have previously been using, signified presently as ψ_1, ψ_2, ..., and a spin function which is α or β. Each function, spatial or spin, has following it parentheses to which an electron is assigned. Then we can write the space-spin orbitals as

$$\Psi_1(1) = \psi_1(1)\alpha(1), \qquad \bar{\Psi}_1(2) = \psi_1(2)\beta(2) \qquad\qquad 5.1\text{-}2$$

However, it is artificial to assume that electron 1 is uniquely assigned to Ψ_1 and electron 2 confined to $\bar{\Psi}_1$. Rather, it is more reasonable to include the alternative assignment

$$\bar{\Psi}_1(1)\Psi_1(2) \qquad\qquad 5.1\text{-}3$$

in linear combination with the original assignment $\Psi_1(1)\bar{\Psi}_1(2)$. The negative linear combination in Eq. 5.1-4 is taken with the philosophy that the total wavefunction should be antisymmetric with respect to exchange of electrons 1 and 2.

$$\Phi = \frac{1}{\sqrt{2}}\left\{\Psi_1(1)\bar{\Psi}_1(2) - \bar{\Psi}_1(1)\Psi_1(2)\right\} \qquad\qquad 5.1\text{-}4$$

Thus, if the operator P signifies permuting electrons 1 and 2, the polyelectron wavefunction Φ is seen to be converted to its negative by permuting with the P operator. Accordingly,

$$P\Phi = \frac{1}{\sqrt{2}}P\left\{\Psi_1(1)\bar{\Psi}_1(2) - \bar{\Psi}_1(1)\Psi_1(2)\right\} = \frac{1}{\sqrt{2}}\left\{\Psi_1(2)\bar{\Psi}_1(1) - \bar{\Psi}_1(2)\Psi_1(1)\right\}$$

$$= \frac{1}{\sqrt{2}}\left\{\bar{\Psi}_1(1)\Psi_1(2) - \Psi_1(1)\bar{\Psi}_1(2)\right\} = -\Phi \qquad\qquad 5.1\text{-}5$$

A wavefunction which is antisymmetric with respect to electron exchange is said to be antisymmetrized.

We note that the wavefunction Φ corresponds to the expansion of a determinant and may be written as

$$\Phi = \frac{1}{\sqrt{2}}\begin{vmatrix} \Psi_1(1) & \bar{\Psi}_1(1) \\ \Psi_1(2) & \bar{\Psi}_1(2) \end{vmatrix}$$

In general antisymmetrized wavefunctions may be written as such determinants where the columns contain the occupied space-spin MOs and each row has a different electron assignment. Hence for a closed-shell system

$$\Phi = \frac{1}{\sqrt{2n!}} \begin{vmatrix} \Psi_1(1)\bar{\Psi}_1(1)\Psi_2(1)\bar{\Psi}_2(1)\Psi_3(1)\bar{\Psi}_3(1)\cdots\bar{\Psi}_n(1) \\ \Psi_1(2)\bar{\Psi}_1(2)\Psi_2(2)\bar{\Psi}_2(2)\Psi_3(2)\bar{\Psi}_3(2)\cdots \\ \Psi_1(3)\cdots \\ \cdot \\ \cdot \\ \cdot \\ \Psi_1(2n)\cdots\cdots\cdots\cdots\cdots\cdots\cdots\cdots\bar{\Psi}_n(2n) \end{vmatrix} \qquad 5.1\text{-}6$$

One feature of the Slater determinant worth noting at this point is the behavior of the function Φ if one attempts to consider an electronic configuration where the same space-spin MO is used more than once, that is, where two electrons are assigned with the same spin to the same space MO. In this case two columns of the Slater determinant become identical and determinant algebra tells us that the polyelectron wavefunction Φ vanishes. Thus Slater determinants enforce the Pauli principle.

A common short and convenient notation allows us to write Eq. 5.1-6 as

$$\Phi = \frac{1}{\sqrt{2n!}}|\Psi_1(1)\bar{\Psi}_1(2)\Psi_2(3)\bar{\Psi}_2(4)\Psi_3(5)\bar{\Psi}_3(6)\cdots\bar{\Psi}_n(2n)| \qquad 5.1\text{-}7$$

Here we write only the diagonal terms of the full Slater determinant with the understanding that the full determinant is implied. Another useful notation is

$$\Phi = \bar{\bar{P}}\Psi_1(1)\bar{\Psi}_1(2)\Psi_2(3)\bar{\Psi}_2(4)\Psi_3(5)\bar{\Psi}_3(6)\cdots\bar{\Psi}_n(2n) \qquad 5.1\text{-}7\text{a}$$

or

$$\Phi = \frac{1}{\sqrt{2n!}}\bar{P}\Psi_1(1)\bar{\Psi}_1(2)\Psi_2(3)\bar{\Psi}_2(4)\Psi_3(5)\bar{\Psi}_3(6)\cdots\bar{\Psi}_n(2n) \qquad 5.1\text{-}7\text{b}$$

or

$$\Phi = \frac{1}{\sqrt{2n!}}\,\Sigma(-1)^P P\Psi_1(1)\bar{\Psi}_1(2)\Psi_2(3)\bar{\Psi}_2(4)\Psi_3(5)\bar{\Psi}_3(6)\cdots\bar{\Psi}_n(2n) \qquad 5.1\text{-}7\text{c}$$

Here the $\bar{\bar{P}}$ operator in Eq. 5.1-7a signifies take all permutations and add these up with appropriate plus or minus signs and then normalize the sum. The $\bar{P}$ operator takes all permutations but omits the normalization. The P operator merely says to permute the product of n space-spin MOs. The summation in Eq. 5.1-7c includes all permutations. Even permutations with two or some even number of permutations are given positive signs [note $(-1)^P$ then is $+1$] and odd permutations are given a minus sign [here $(-1)^P$ is -1]. For example,

$$P\Psi_1(1)\bar{\Psi}_2(2) = -\Psi_1(2)\bar{\Psi}_2(1) = -\bar{\Psi}_2(1)\Psi_1(2) \qquad 5.1\text{-}8$$

We note that it does not make any difference if we permute the electrons or instead keep the electron assignments in the same order and, instead, permute the space-spin MOs.

Finally, thinking about the full secular determinant in Eq. 5.1-6, we realize that the determinant is, by definition, just the same sum of n-fold products as in Eq. 5.1-7. Thus the definition of a determinant tells us to select one space-spin MO with its electron assignment from column 1, to multiply this by an MO with electron assignment selected from column 2, and to multiply this from such a function selected from column 3, etc. In selecting functions from each column to make up a product we must not use the same row twice and we must affix a plus or minus sign depending on whether the product is even or odd (i.e., whether the permutation is even or odd). We then add together all such permutations and normalize to get Φ. Our shorthand notation of Eq. 5.1-7 explicitly gives the zeroth permutation.

5.2   Energy of a Single Slater Determinantal Wavefunction

To obtain the energy of a single Slater determinant including effects due to mutual electron–electron repulsion we have to consider two operators and derived integrals. One operator we have already discussed, namely $\mathfrak{IC}_i$, defined by

$$I_k = \int \Psi_k(i)\,\mathfrak{IC}\,\Psi_k(i)\,d\tau = \int \psi_k(i)\,\mathfrak{IC}\,\psi_k(i)\,d\tau \qquad 5.2\text{-}1$$

gives the energy I_k of electron i in MO k as a result of its kinetic energy and also potential energy due to attraction by molecular nuclei.

A new operator $\mathcal{G}_{ij} = e^2/r_{ij}$ represents the potential energy resulting from mutual repulsion of two electrons i and j. Using this we can write two-electron integrals such as

$$G_{klkl}{}^{\text{MO}} = \int \psi_k(i)\psi_l(j)\mathcal{G}_{ij}\psi_k(i)\psi_l(j)\,d\tau \qquad 5.2\text{-}2$$

which represents the mutual repulsion of electrons i and j with electron i

assigned to space MO k and electron j assigned to space MO l:

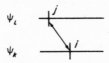

Note that here spin is not involved. For example, if k and l were to apply to MOs 1 and 2 of ethylene, the term G_{1212} then gives us the energy of repulsion between an electron in the bonding MO and an electron in the antibonding MO (i.e., the repulsion between the two one-electron clouds in Fig. 5.2-A, dotted area and hatched area). Note that the subscripts in G_{klkl} are in the same order as in the terms under the integral in Eq. 5.2-2 where the MOs are listed in the order of functions of electrons i, j, i, j sequentially.

The total energy operator is taken as

$$\mathfrak{F} = \mathfrak{IC} + \mathfrak{G} \qquad\qquad 5.2\text{-}3$$

We can now proceed to obtain the energy of a Slater determinant including both one-electron and two-electron terms. Thus

$$\Phi = \frac{1}{\sqrt{2n!}}\, \bar{P}\Psi_1(1)\bar{\Psi}_1(2)\Psi_2(3)\cdots\bar{\Psi}_n(2n) \qquad\qquad 5.2\text{-}4$$

and the integrated form of the Schrödinger equation gives us

$$E = \frac{1}{2n!}\int \bar{P}'\Psi_1(1)\bar{\Psi}_1(2)\Psi_2(3)\cdots\bar{\Psi}_n(2n)\mathfrak{F}\bar{P}''\Psi_1(1)\bar{\Psi}_1(2)\Psi_2(3)\cdots\bar{\Psi}_n(2n)\ d\tau$$

$$5.2\text{-}5$$

The designation of the permutation operators by a prime (′) and a double prime (″) merely indicates that each permutation is independent of the other.

Equation 5.2-5 can be simplified since it is not necessary to permute both the MO products before and after the operator; it is sufficient to permute only one. To show this, let us premultiply the integral in 5.2-5 by a permutation operator P_q so designed that each permutation resulting from

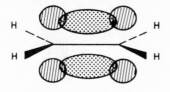

F IG. 5.2-A

$\bar{P}'$ is reversed to restore the original product of MOs. We note that P_q affects the electron assignment in both the MO products before and after $\mathfrak{F}$ but permuting all the electrons does not change the value of the integral.

We also note that $\bar{P}'$ has the $2n!$ terms and after this reverse permutation we have $2n!$ identical permutations preceding $\mathfrak{F}$. Thus the normalization factor of $1/2n!$ in 5.2-6a is canceled in 5.2-6b

$$E = \frac{1}{2n!} P_q \int \bar{P}' \Psi_1(1) \bar{\Psi}_1(2) \Psi_2(3) \bar{\Psi}_2(4) \cdots \bar{\Psi}_n(2n)$$

$$\times \mathfrak{F} \bar{P}'' \Psi_1(1) \bar{\Psi}_1(2) \Psi_2(3) \bar{\Psi}_2(4) \cdots \bar{\Psi}_n(2n) \, d\tau \qquad \text{5.2-6a}$$

$$= \int \Psi_1(1) \bar{\Psi}_1(2) \Psi_2(3) \bar{\Psi}_2(4) \cdots \bar{\Psi}_n(2n)$$

$$\times \mathfrak{F} P_q \bar{P}'' \Psi_1(1) \bar{\Psi}_1(2) \Psi_2(3) \bar{\Psi}_2(4) \cdots \bar{\Psi}_n(2n) \, d\tau \qquad \text{5.2-6b}$$

$$= \int \Psi_1(1) \bar{\Psi}_1(2) \Psi_2(3) \bar{\Psi}_2(4) \cdots \Psi_n(2n)$$

$$\times \mathfrak{F} \bar{P} \Psi_1(1) \bar{\Psi}_1(2) \Psi_2(3) \bar{\Psi}_2(4) \cdots \bar{\Psi}_n(2n) \, d\tau \qquad \text{5.2-6c}$$

But $P_q \bar{P}'$ is just equivalent to an ordinary unnormalized sum of permutations $\bar{P}$ since all permutations are included. Equation 5.2-6c then gives a very convenient form for writing this and similar integrals involving two Slater determinantal wavefunctions. All we must do is to include all permutations in the second term but without any normalization needed.

To evaluate 5.2-6c it is helpful to consider the $\mathfrak{K}$ and $\mathfrak{G}$ components of $\mathfrak{F} = \mathfrak{K} + \mathfrak{G}$ separately, or,

$$E = E_H + E_G \qquad \text{5.2-7}$$

The $\mathfrak{K}$ component includes the one-electron energy effects, kinetic and potential energy, for all electrons. That is,

$$\mathfrak{K} = \mathfrak{K}_1 + \mathfrak{K}_2 + \mathfrak{K}_3 + \cdots + \mathfrak{K}_{2n} \qquad \text{5.2-8}$$

where $\mathfrak{K}_1$ is the operator for electron 1, $\mathfrak{K}_2$ for electron 2, and so on. Thus, the one-electron component of the energy E is given by

$$E_H = \int \Psi_1(1) \bar{\Psi}_1(2) \Psi_2(3) \bar{\Psi}_2(4) \cdots \bar{\Psi}_n(2n)$$

$$\times [\mathfrak{K}_1 + \mathfrak{K}_2 + \mathfrak{K}_3 + \cdots] \bar{P} \Psi_1(1) \bar{\Psi}_1(2) \Psi_2(3) \bar{\Psi}_2(4) \cdots \bar{\Psi}_n(2n) \, d\tau$$

$$\text{5.2-9}$$

Similarly, the two-electron component of the energy

$$E_G = \int \Psi_1(1)\bar\Psi_1(2)\Psi_2(3)\bar\Psi_2(4)\cdots\bar\Psi_n(2n)$$

$$\times\ \mathcal{G}\bar P\Psi_1(1)\bar\Psi_1(2)\Psi_2(3)\bar\Psi_2(4)\cdots\bar\Psi_n(2n)\ d\tau \qquad 5.2\text{-}10$$

Here the two-electron operator represents the sum of all sets of repulsions between two electrons and thus

$$\mathcal{G} = \mathcal{G}_{12} + \mathcal{G}_{13} + \mathcal{G}_{14} + \cdots + \mathcal{G}_{23} + \mathcal{G}_{24} + \cdots + \mathcal{G}_{34} + \cdots \qquad 5.2\text{-}11$$

Turning first to the one-electron integral, we note that this is the sum of separate integrals, the first using $\mathcal{3C}_1$, the second using $\mathcal{3C}_2$, the third using $\mathcal{3C}_3$, and so on. We also note that each of these integrals can be broken up into the product of integrals, each one of which is a function of only one electron. For example, if we consider the integral with the energy operator for electron 1, we find this to be

$$E_{H1} = \int_1 \Psi_1(1)\mathcal{3C}_1\Psi_1(1)\ d\tau_1 \int_2 \bar\Psi_1(2)\bar\Psi_1(2)\ d\tau_2$$

$$\times \int_3 \Psi_2(3)\Psi_2(3)\ d\tau_3 \int_4 \bar\Psi_2(4)\bar\Psi_2(4)\ d\tau_4\cdots d\tau_{2n} \qquad 5.2\text{-}12$$

This has assumed only the zeroth permutation of the operator $\bar P$, and we will have to justify the lack of permutation in arriving at this result. However, first let us evaluate the integral product which gives E_{H1} (note Eq. 5.2-12). In this product we see that each integration involves functions of just one electron; in fact, it was this which allowed dissection of the original integration over all electrons into the simpler form above. We note that except for the first integration, all the integrals have the value of 1 by virtue of normalization. The first integral is just I_1 (note Eq. 5.2-1) which is the one-electron energy of MO 1:

$$E_{H1} = I_1 \qquad 5.2\text{-}13$$

In arriving at this equation we have assumed that all permutations except for the original one (i.e., the zeroth one) lead to zero integrals. This can be seen in the following. Thus, in Eq. 5.2-9, if we still retain consideration of only $\mathcal{3C}_1$ and permute any two space-spin MOs, we obtain a vanishing integral. If the two MOs permuted are Ψ_1 and $\bar\Psi_1$, we obtain the integral in Eq. 5.2-14. This is zero due to two integrals being zero, namely the integral involving electron 1 and also the integral involving electron 2. The two integrals vanish due to spin orthogonality.* The integrals of Eq. 5.2-9 which are not permuted are still unity and are not explicitly written out in the following:

$$\int_1 \Psi_1(1)\mathcal{3C}_1\bar{\Psi}_1(1) \; d\tau_1 \int_2 \Psi_1(2)\bar{\Psi}_1(2) \; d\tau_2$$

$$= \int_1 \psi_1(1)\mathcal{3C}_1\psi_1(1) \; d\tau_1 \int_2 \psi_1(2)\psi_1(2) \; d\tau_2$$

$$\times \int_1 \alpha(1)\beta(1) \; d\tau_1 \int_2 \alpha(2)\beta(2) \; d\tau_2 = 0 \qquad 5.2\text{-}14$$

If we try to avoid the spin orthogonality problem and again try to permute terms in Eq. 5.2-9, we might try exchanging Ψ_1 and Ψ_2. Again considering only the $\mathcal{3C}_1$ operator, we obtain from 5.2-9 terms which are zero:

$$\int_1 \Psi_1(1)\mathcal{3C}_1\Psi_2(1) \; d\tau_1 \int_2 \bar{\Psi}_1(2)\bar{\Psi}_1(2) \; d\tau_2 \int_3 \Psi_2(3)\Psi_1(3) \; d\tau_3 = 0 \quad 5.2\text{-}15$$

In this case it can be seen that the permuted integral product vanishes due to the third integral which is zero as a result of spatial orthogonality. Also the first integral in the product will be zero if ψ_1 and ψ_2 are eigenfunctions of the one-electron operator $\mathcal{3C}_1$.

In the same fashion we can evaluate the part of the energy deriving from the other one-electron operators, namely $\mathcal{3C}_2$, $\mathcal{3C}_3$, etc., and conclude that each of these affords a contribution of I_k where k is the MO containing the electron of the one-electron operator. Thus,

$$E_H = \sum_k n_k I_k \qquad 5.2\text{-}16$$

Here n_k is the number of electrons assigned to MO k. We next need to evaluate E_G (note Eq. 5.2-10). In evaluating 5.2-10 we recognize that the operator is the summation of all two-electron operators in Eq. 5.2-11. Without any permutation, we find that each use of an operator $\mathcal{G}_{ij}$ results in a two-electron integral of the type in Eq. 5.2-2, and this is multiplied by a product of integrals that are all unity by normalization. For example,

---

* Integrals of products of spin functions are unity when the two spin functions under the integral sign are the same and zero when they are not. For example

$$\int_1 \alpha^2(1) \; d\tau = 1 \quad \text{while} \quad \int_1 \alpha(1)\beta(1) \; d\tau = 0.$$

use of the $\mathcal{G}_{13}$ component gives

$$E_{G_{13}}{}^0 = \int_{1,3} \Psi_1(1)\Psi_2(3)\mathcal{G}_{13}\Psi_1(1)\Psi_2(3)\ d\tau_1\,d\tau_3 \int_2 \bar{\Psi}_1{}^2(2)\ d\tau_2 \int_4 \bar{\Psi}_2{}^2(4)\ d\tau_4$$

$$= G_{1212}{}^{\mathrm{MO}} \qquad\qquad\qquad 5.2\text{-}17$$

Using all components of the operator we see that we will obtain one G_{1111}, four G_{1212}, one G_{2222}, and so on, terms. This can be formulated as

$$E_G{}^0 = \sum_{k,l} n_{kl} G_{klkl}{}^{\mathrm{MO}} \qquad\qquad 5.2\text{-}18$$

where this gives the terms arising without permutation. Here n_{kl} is the number of possible pairs of electrons with one in MO k and the other in MO l.

We now have to consider terms arising from 5.2-10 if we include permutation of the electrons in the space-spin MOs following the repulsion operator. Perfectly equivalent to permuting the electrons is permutation of the MOs, thus keeping the terms in order of increasing electron number, and this is used here. First, we can see that any permutation of space-spin MOs of different spin leads to spin orthogonality. Second, we can also see that any two space-spin MOs permuted must be those containing the two electrons of the operator. Otherwise, we will obtain vanishing orthogonality integrals in the product. The preceding leads to terms such as

$$E_{G_{13}}{}^P = -\int_{1,3} \Psi_1(1)\Psi_2(3)\mathcal{G}_{13}\Psi_2(1)\Psi_1(3)\ d\tau_1\,d\tau_3 \int_2 \bar{\Psi}_1{}^2(2)\ d\tau_2 \int_4 \bar{\Psi}_2{}^2(4)\ d\tau_4 \cdots$$

$$= -\int_{1,3} \psi_1(1)\psi_2(3)\mathcal{G}_{13}\psi_2(1)\psi_1(3)\ d\tau_1\,d\tau_3$$

$$= -G_{1221}{}^{\mathrm{MO}} \qquad\qquad\qquad 5.2\text{-}19$$

The negative sign results from use of the permutation operator. The total energetic contribution from such permuted terms is

$$E_G{}^P = -\sum_{k,l} m_{kl} G_{kllk} \qquad\qquad 5.2\text{-}20$$

Here m_{kl} is the number of possible pairs of electrons with one in MO k and the other in MO l; in contrast to n_{kl}, however, the two electrons constituting each pair must have the same spin.

Thus we have three types of contributions to the total electronic energy. One is just the sum of the one-electron energies (note Eq. 5.2-16). If the MOs are Hückel MOs, then the energy will be the same as that which we have considered thus far.

The second contribution is a sum of electron repulsions as given in Eq.

5.2-18. This is destabilizing and derives from mutual electron–electron repulsion. We can obtain this term very simply from consideration of all repelling pairs.

The third contribution is given in Eq. 5.2-20 and this is seen to be stabilizing. The integrals are termed exchange integrals and the stabilization derives from the ability of electrons of like spin to permute with the result of diminution of electron repulsion.

The three energy contributions are shown schematically in Fig. 5.2-B for a four-MO system (e.g., butadiene).

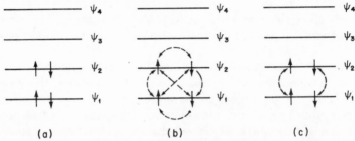

(a)   (b)   (c)

FIG. 5.2.B. Schematic representation of contributions to the energy of a closed-shell system. (a) One-electron terms, contribution $2I_1 + 2I_2$; (b) electron–electron repulsion, contribution $G_{1111} + 4G_{1212} + G_{2222}$; (c) electron–electron exchange terms, $2G_{1221}$.

We can now combine all the contributions to the energy of a Slater determinant and obtain the general expression

$$E = \sum_k n_k I_k + \sum_{k \leq l} n_{kl} G_{klkl} - \sum_{k \leq l} m_{kl} G_{kllk} \qquad \text{5.2-21}$$

It should be recognized that this expression gives the energy of any Slater determinant, independent of whether or not it is a closed-shell system. However, we have not yet discussed the significance of open-shell Slater determinants and these are not always proper wavefunctions.

Since we are presently dealing with closed-shell systems, we can put in definite values of the n_k's, the n_{kl}'s, and the m_{kl}'s. It can be seen that n_k is 2 for each doubly occupied MO. Second, n_{kl} will be 1 for each G_{kkkk}-type term and will be 4 (note Fig. 5.2-B for example) for each G_{klkl} term where k and l are different MOs. Similarly, m_{kl} will be 2 for each G_{kllk}-type term where k and l are different but m_{kl} is zero for $k = l$ since of necessity the electrons in this one MO (i.e., termed k or l) are of opposite spin and m_{kl} gives contributions only for pairs of electrons of the same spin.

We can write the total energy for a closed-shell system as

$$E = 2 \sum_k^{occ} I_k + \sum_k^{occ} G_{kkkk} + 4 \sum_{k<l}^{occ} G_{klkl} - 2 \sum_{k<l}^{occ} G_{kllk} \qquad \text{5.2-22}$$

By keeping $k < l$ in these summations, we assure ourselves that the same electron–electron repulsion term is not counted twice. However, we can rewrite the last two terms in 5.2-22 allowing all values of occupied MOs for both k and l. But, then we have to divide these summations by 2 since each term is duplicated. Hence

$$E = 2 \sum_k^{occ} I_k + \sum_k^{occ} G_{kkkk} + 2 \sum_{k \neq l}^{occ} G_{klkl} - \sum_{k \neq l}^{occ} G_{kllk} \qquad 5.2\text{-}23$$

If in the last two terms of 5.2-23 we were to omit the restriction that $k \neq l$, the last term would be increased negatively by a summation of G_{kkkk} terms and the next to last term would be increased positively by twice a summation of the same terms. Thus, we can omit the $k \neq l$ restriction in the last two summations by just compensating and not including the second term in 5.2-23. Thus

$$E = 2 \sum_k^{occ} I_k + 2 \sum_{k,l}^{occ} G_{klkl} - \sum_{k,l}^{occ} G_{kllk} \qquad 5.2\text{-}24$$

We note that these summations are over all occupied MOs.

In considering equations such as 5.2-21 and 5.2-24, we have to recognize that these merely give the total electronic energy of the MOs we are using. If these are Hückel MOs, for example, the energies derived will be correct for these orbitals but will be higher than if we use better molecular orbitals. This point has to be considered subsequently.

We still have not discussed the nature of the terms in the summations except by definition in Eq. 5.2-2. Terms of the type G_{klkl} have been noted to derive from electron–electron repulsion with one electron in MO k and one in MO l. However, these integrals can be evaluated further as in the next section.

Turning to one other point, we find it is convenient to define single-electron (i.e., MO) energies as

$$\epsilon_k = I_k + 2 \sum_l G_{klkl}{}^{MO} - \sum_l G_{kllk}{}^{MO} \qquad 5.2\text{-}25$$

This gives the energy of a single electron in MO k as a consequence of its one-electron contribution (i.e., the I_k) and also including its repulsion by the other electrons and taking into account stabilization by exchange with electrons of like spin. But note that the total electronic energy is not the sum of these one-electron energies. Thus,

$$E \neq \sum_k^{occ} 2\epsilon_k = 2 \sum_k^{occ} I_k + 4 \sum_{k,l}^{occ} G_{klkl} - 2 \sum_{k,l}^{occ} G_{kllk} \qquad 5.2\text{-}26$$

since this would include each electron repulsion and exchange term twice as needed. (Compare with Eq. 5.2-24.) Hence Eq. 5.2-25 gives only the energy experienced by each electron as a result of interactions with all of the other electrons. By adding such energies together we thus are including each electron–electron repulsion and each electron–electron exchange twice.

5.3   Evaluation of MO Repulsion Integrals

If one were dealing with the butadiene problem, one would obtain terms such as $G_{1111}{}^{MO}$, $G_{1212}{}^{MO}$, $G_{2222}{}^{MO}$, and $G_{1221}{}^{MO}$. These may or may not be explicitly labeled with a superscript "MO," but they do represent interaction between MOs. Since each molecule requires its own set of such integrals it is not realistic to have these tabulated and available.

However, such integrals can be evaluated in terms of more available quantities. To do this we write down the definition of a general MO repulsion integral (e.g., G_{klmn}) and substitute in the LCAO-MO expansion for each MO in the integral.

$$G_{klmn} = \int_{1,2} \psi_k(1)\psi_l(2)\mathsf{G}_{12}\psi_m(1)\psi_n(2)\ d\tau_1\ d\tau_2 \qquad \text{5.3-1a}$$

$$= \int \sum_r c_{rk}\chi_r(1) \sum_s c_{sl}\chi_s(2)\mathsf{G} \sum_t c_{tm}\chi_t(1) \sum_u c_{un}\chi_u(2)\ d\tau_1\ d\tau_2 \qquad \text{5.3-1b}$$

$$= \sum_{rstu} c_{rk}c_{sl}c_{tm}c_{un} \int \chi_r(1)\chi_s(2)\mathsf{G}\chi_t(1)\chi_u(2)\ d\tau_1\ d\tau_2 \qquad \text{5.3-1c}$$

The last integral involves atomic rather than molecular orbitals. There are a number of conventions for writing such repulsion integrals. Among these are $(rs|\mathsf{G}|tu)$, $(rt\mid su)$, and $G_{rstu}{}^{AO}$. In the first of these, the orbital subscripts are kept in the same order as in the integral (i.e., electron 1, electron 2, electron 1, electron 2). In the second notation, the subscripts before the vertical bar refer to functions of electron 1 and those after the bar refer to electron 2.

This integral often has been simplified by an approximation termed neglect of differential overlap, or zero differential overlap. This is based on the idea that $\chi_r\chi_t$ will be vanishingly small unless $r = t$. One can see that if these orbitals are not identical, then any volume element dv present in one where the value of the atomic orbital is appreciable will be in a region in space where the value of the other orbital is small (Fig. 5.3-A). This is a more restrictive assumption than the usual neglect of overlap

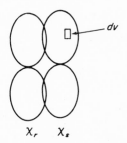

$$\chi_r \quad \chi_s$$

Fig. 5.3-A

where merely the entire overlap integral is taken to be zero. Here we are assuming that the orbital product is zero even before integration.

If we assume zero differential overlap (ZDO), then Eq. 5.3-1c simplifies considerably. For the integral not to vanish, it is then required that $t = r$ and that $u = s$. We then obtain

$$G_{klmn}{}^{MO} = \sum_{r,s} c_{rk}c_{sl}c_{rm}c_{sn} \int_{1,2} \chi_r(1)\chi_s(2)\mathcal{G}\chi_r(1)\chi_s(2) \; d\tau_1 \, d\tau_2 \qquad 5.3\text{-}2a$$

$$= \sum_{r,s} c_{rk}c_{sl}c_{rm}c_{sn} \int_{1,2} \chi_r{}^2(1)\mathcal{G}\chi_s{}^2(2) \; d\tau_1 \, d\tau_2 \qquad 5.3\text{-}2b$$

The integral in Eq. 5.3-2b can be seen to be the atomic orbital analog of the MO repulsion integral in Eq. 5.2-2. Presently the integral represents the energy of repulsion of electron 1 in atomic orbital χ_r with electron 2 in atomic orbital χ_s. Thus the term $\chi_r{}^2(1)$ gives the electron density as a function of position in space of electron 1. Similarly, $\chi_s{}^2(2)$ gives the electron density of electron 2 as a function of its coordinates. For any two points in space, one having electron 1 and the other giving the position of electron 2, the product

$$\chi_r{}^2(1) \, (e^2/r_{12}) \chi_s{}^2(2)$$

gives the electrostatic repulsion energy between the two electron densities. Then the integral in Eq. 5.3-2b merely affords the total repulsion when one integrates over all space. Finally, a shorthand abbreviation for the repulsion integral is just γ_{rs} so that Eq. 5.3-2b becomes

$$G_{klmn} = \sum_{r,s} c_{rk}c_{sl}c_{rm}c_{sn}\gamma_{rs} \qquad 5.3\text{-}3$$

If we define a quantity $\omega_{r,km} = c_{rk}c_{rm}$, then we can rewrite 5.3-1 conveniently in matrix form as a vector–matrix–vector triple product:

$$G_{klmn}{}^{MO} = \sum_{rs} \omega_{r,km}\gamma_{rs}\omega_{s,ln} = \tilde{\boldsymbol{\omega}}_{km}\boldsymbol{\Gamma}\boldsymbol{\omega}_{ln} \qquad 5.3\text{-}4$$

where

$$\widetilde{\omega}_{km} = \begin{bmatrix} c_{1k}c_{1m} & c_{2k}c_{2m} & c_{3k}c_{3m} & c_{4k}c_{4m} & \cdots \end{bmatrix}$$

and ω_{ln} is a similar vector whose elements are the same type of products of the LCAO-MO coefficients deriving from MOs l and n. Thus, if k and m were equal in ω_{km} or l and n were equal in ω_{ln}, the elements of that vector would be electron density terms (i.e., just the LCAO-MO coefficients squared). As it is, the elements of the omega vectors are products of coefficients for a given atom but derived from two different MOs. Finally, the γ_{rs} terms used, which are noted above to be the energy of repulsion between two electrons with one in atomic orbital r and the other in atomic orbital s, depend on the elements bearing the two orbitals, the nature of the orbitals, and their distance apart.

There is a very simple way of evaluating the triple vector–matrix–vector product in Eq. 5.3-4. This can be demonstrated most easily for a specific case. For example, suppose we wish to obtain $G_{1221}{}^{MO}$, that is $[12 \mid 12]^{MO}$, for ethylene. For this we need the vector

$$\widetilde{\omega}_{12} = \begin{bmatrix} c_{11}c_{12} & c_{21}c_{22} \end{bmatrix} = \begin{bmatrix} \tfrac{1}{2} & -\tfrac{1}{2} \end{bmatrix} \qquad\qquad 5.3\text{-}5$$

as well as its transpose. To solve for $G_{1221}{}^{MO} = [12 \mid 12]^{MO}$ we need $\widetilde{\omega}_{12}\,\Gamma\,\omega_{12}$. This is written out as

$$\begin{bmatrix} \tfrac{1}{2} & -\tfrac{1}{2} \end{bmatrix} \begin{bmatrix} \gamma_{11} & \gamma_{12} \\ \gamma_{21} & \gamma_{22} \end{bmatrix} \begin{bmatrix} \tfrac{1}{2} \\ -\tfrac{1}{2} \end{bmatrix} = G_{1221}{}^{MO} \qquad\qquad 5.3\text{-}6$$

The product of the vector–matrix multiplication (i.e., the first two terms) is a vector and multiplication by the final vector then gives a scalar value for the repulsion integral in terms of the AO repulsion integrals. However, one can do this more simply by just remembering that post-multiplication by a vector is equivalent to taking a linear combination of the columns and premultiplication is equivalent to taking a linear combination of the rows. Thus, we use the elements of the postmultiplying vector and label the columns of the Γ matrix in order with these elements. Each column is multiplied through by its label. Similarly, we use the elements of the pre-multiplying row vector to label and multiply the rows. After these multiplications one adds up all elements. This process affords the same scalar result that formal triple matrix multiplication would. However, it allows one to see, prior to the actual operation, just how many times each repulsion integral in the Γ matrix is used in the final result.

This is illustrated in 5.3-7*:

$$
\begin{array}{cc} +\tfrac{1}{2} & -\tfrac{1}{2} \end{array}
$$

$$
\begin{array}{c} +\tfrac{1}{2} \\ -\tfrac{1}{2} \end{array}
\begin{bmatrix} \gamma_{11} & \gamma_{12} \\ \gamma_{21} & \gamma_{22} \end{bmatrix}
\equiv
\begin{pmatrix} S \\ U \\ M \end{pmatrix}
\begin{bmatrix} +\tfrac{1}{4}\gamma_{11} & -\tfrac{1}{4}\gamma_{12} \\ -\tfrac{1}{4}\gamma_{21} & +\tfrac{1}{4}\gamma_{22} \end{bmatrix}
$$

$$
= \tfrac{1}{4}\gamma_{11} - \tfrac{1}{4}\gamma_{12} - \tfrac{1}{4}\gamma_{21} + \tfrac{1}{4}\gamma_{22} \qquad 5.3\text{-}7
$$

For illustration purposes all of the terms resulting are kept separate. But we recognize that $\gamma_{11} = \gamma_{22}$ are identical as are γ_{12} and γ_{21}. The former are just repulsion integrals indicating the energy raising due to two electrons in a single p orbital and the latter are comparable terms but where the electrons are in adjacent (i.e., vicinal) p orbitals. Thus, we can see immediately from simple inspection of the left matrix in 5.3-7 that each diagonal term is taken $(\tfrac{1}{2})(\tfrac{1}{2})$ times and that there are two equal such terms. We can see each off-diagonal term is multiplied by a positive term (i.e., $+\tfrac{1}{2}$) and by a negative term (i.e., $-\tfrac{1}{2}$) and that we have a total of two off-diagonal terms, each then multiplied $(\tfrac{1}{2})(-\tfrac{1}{2})$ times. We thus come out with $0.5\gamma_{11} - 0.5\gamma_{12}$ as the value of $G_{1221}{}^{MO}$. For larger systems the method proves especially useful, since often it is quickly possible to tell when an MO repulsion integral vanishes and it generally is easy to total up the number of each type of atomic orbital repulsion integrals. It turns out to be general that when the symmetry of the pre- and postmultiplying vectors differs, the integral becomes zero, and consideration of the multiplication of columns and rows by elements derived from omega vectors of different symmetries reveals that a cancellation of terms will occur.

5.4 Energy of a Slater Determinant for a Closed Shell in Terms of Atomic Orbital Integrals

We have now obtained in Eq. 5.2-24 an expression for the total electronic energy of a closed-shell system. However, the result is in terms of MO repulsion and exchange integrals of the type $G_{klkl}{}^{MO}$ and $G_{kllk}{}^{MO}$. We now would like to obtain the electronic energy in terms of atomic orbital repul-

* The

$$
\begin{pmatrix} S \\ U \\ M \end{pmatrix}
$$

operator indicates the summation of all matrix elements.

sion integrals. We need to expand I_k in terms of atomic orbital integrals prior to completing this. We see that

$$I_k = \int \psi_k(1)\mathfrak{K}\psi_k(1)\ d\tau = \int \sum_r c_{rk}\chi_r(1)\mathfrak{K} \sum_t c_{tk}\chi_t(1)\ d\tau$$

$$= \sum_{r,t} c_{rk}c_{tk} \int \chi_r(1)\mathfrak{K}\chi_t(1)\ d\tau = \sum_{r,t} c_{rk}c_{tk}H_{rt} \qquad \text{5.4-1}$$

If we now take Eq. 5.2-24 and substitute for the MO one-electron energy terms (using 5.4-1) and also for the MO repulsion integrals (i.e., using 5.3-1), we obtain

$$E = 2 \sum_{rt,k} c_{rk}c_{tk}H_{rt} + \sum_{rstu,kl} c_{rk}c_{sl}c_{tk}c_{ul}(2G_{rstu}{}^{AO} - G_{rsut}{}^{AO}) \qquad \text{5.4-2a}$$

or equivalently

$$E = 2 \sum_{rt,k} c_{rk}c_{tk}H_{rt} + \sum_{rstu,kl} (2c_{rk}c_{sl}c_{tk}c_{ul} - c_{rk}c_{sl}c_{tl}c_{uk})G_{rstu}{}^{AO} \qquad \text{5.4-2b}$$

In Eq. 5.4-2b we have exchanged the subscripts t and u in the last summation of 5.4-2a. This is acceptable since we are summing over all values (i.e., over all atoms t and u) and the letter used to designate an atom is arbitrary.

We might assume zero differential overlap now. Then $G_{rstu}{}^{AO}$ vanishes unless $t = r$ and $u = s$; this deletes all other terms in the summation in 5.4-2b to give

$$E = 2 \sum_{rt,k} c_{rk}c_{tk}H_{rt} + \sum_{rs,kl} (2c_{rk}{}^2c_{sl}{}^2 - c_{rk}c_{sl}c_{rl}c_{sk})\gamma_{rs} \qquad \text{5.4-3a}$$

where $\gamma_{rs} = G_{rsrs}{}^{AO}$.

If we designate one-electron bond orders by $p_{rt,k} = c_{rk}c_{tk}$, one-electron densities by $q_{rk} = c_{rk}{}^2$, and again use the notation that $\omega_{r,kl} = c_{rk}c_{rl}$, we obtain

$$E = 2 \sum_{rt,k} p_{rt,k}H_{rt}{}^{AO} + 2 \sum_{rs,kl} q_{rk}q_{sl}\gamma_{rs} - \sum_{rs,kl} \omega_{r,kl}\omega_{s,kl}\gamma_{rs} \qquad \text{5.4-3b}$$

Here we see that the energy of a Slater determinant includes the one-electron energies of Hückel theory in the first term (i.e., remember our treatment of bond order contributions to energy). The second term is a pure Coulombic repulsion term involving two-electron densities, q_{rk} and q_{sl}, repelling one another with an energy of γ_{rs} per unit electron density in each atomic orbital. The final term is stabilizing and derives from our ability to exchange (permute) two electrons of the same spin and thus minimize electron repulsion by allowing an electron at atom r and one at atom s to avoid one another. The terms $\omega_{r,kl}$ and $\omega_{s,kl}$ each give a measure of the exchange occurring.

5.5 Minimization of the Energy of a Slater Determinant; Roothaan's SCF Equations

One can minimize the energy E as given in Eq. 5.4-2a with respect to the LCAO-MO coefficients used. In doing this we have to maintain the requirement for orthonormality of the eigenfunctions. The result given is that derived by Roothaan[1]:

$$\sum_t \{[H_{rt} + \sum_{su,l} c_{sl}c_{ul}(2G_{rstu}{}^{AO} - G_{rsut}{}^{AO})] - \epsilon S_{rt}\}c_{rt} = 0 \qquad 5.5\text{-}1a$$

which holds for $r = 1, 2, 3, \ldots, n$. It is seen that Eq. 5.5-1a is reminiscent of our usual secular equations except that the quantity in the brackets replaces the usual H_{rt}; actually H_{rt} has an addition to it of

$$\sum_{su,l} c_{sl}c_{ul}(2G_{rstu}{}^{AO} - G_{rsut}{}^{AO})$$

We define the entire quantity as

$$F_{rt}{}^{AO} = H_{rt}{}^{AO} + \sum_{su,l} c_{sl}c_{ul}(2G_{rstu}{}^{AO} - G_{rsut}{}^{AO}) \qquad 5.5\text{-}2$$

which then gives

$$\sum_t (F_{rt}{}^{AO} - \epsilon S_{rt})c_{rt} = 0 \qquad 5.5\text{-}1b$$

Thus to solve the secular equations we would proceed in the usual manner, here diagonalizing the F matrix, except that the matrix elements, that is, the F_{rt}'s, need to be obtained from Eq. 5.5-2. The matrix elements are not the simple $H_{rt}{}^{AO}$'s (i.e., α's and β's) which we have previously been using. Rather they are additional terms which can be seen to include electron repulsion and exchange effects on the electron. Unfortunately, evaluation of each F_{rt} element requires the LCAO-MO coefficients. Since these are not available until after diagonalization the best we can do is to start with an approximation to these (e.g., with Hückel coefficients) and then use the resulting coefficients for a second iteration. The process then is repeated until the MO energies and coefficients converge to self-consistency.

Finally, let us write the total energy in terms of the matrix elements we have defined. Thus, we can factor Eq. 5.4-2a to give

$$E = \sum_{rt,k} c_{rk}c_{tk}[2H_{rt} + \sum_{su,l} c_{sl}c_{ul}(2G_{rstu}{}^{AO} - G_{rsut}{}^{AO})] \qquad 5.5\text{-}3$$

We can substitute the definition of $F_{rt}{}^{AO}$ (note Eq. 5.5-2) into this to give

$$E = \sum_{rt,k} c_{rk}c_{tk}[H_{rt}{}^{AO} + F_{rt}{}^{AO}]$$

5.5-4

Then we use this to evaluate the energy obtained at the end of each iteration.

5.6 Pople's SCF Equations

Thus far, the Roothaan SCF equations have not assumed zero differential overlap. If we wish to introduce this assumption (ZDO), it is convenient to consider the effect of the assumption on the value of F_{rt} in Eq. 5.5-2 in two separate cases.

In the first case where we are dealing with diagonal matrix elements (i.e., F_{rr}), $t = r$. Then the first term of 5.5-2 becomes $H_{rr}{}^{AO}$. Additionally, the summation

$$2 \sum_{su,l} c_{sl}c_{ul}G_{rstu}{}^{AO} \qquad \text{becomes} \qquad 2 \sum_{s,l} c_{sl}{}^2 G_{rsrs}{}^{AO}$$

5.6-1

since t must equal r and $u = s$ by our ZDO assumption. Finally the summation

$$-\sum_{su,l} c_{sl}c_{ul}G_{rsut}{}^{AO} \qquad \text{becomes} \qquad -\sum_{l} c_{rl}{}^2 G_{rrrr}$$

5.6-2

since here $u = r$, $s = t$ from ZDO, and also $t = r$ by our initial assumption of the diagonal nature of F_{rt}. Thus

$$F_{rr}{}^{AO} = H_{rr}{}^{AO} + 2 \sum_{s,l} c_{sl}{}^2 \gamma_{rs} - \sum_{l} c_{rl}{}^2 \gamma_{rr}$$

5.6-3

where $\gamma_{rs} = G_{rsrs}$ and $\gamma_{rr} = G_{rrrr}$. We can simplify this further using the definition of electron density and separating out the term for $s = r$ from the second term; thus

$$F_{rr}{}^{AO} = H_{rr}{}^{AO} + \sum_{s \neq r} q_s{}' \gamma_{rs} + \tfrac{1}{2} q_r{}' \gamma_{rr}$$

5.6-4

where $q_r{}'$ is the total electron density at atom r (i.e., $\sum_l n_l q_{rl}$).

For off-diagonal terms F_{rt} is again obtained from Eq. 5.5-2. The first term is now $H_{rt}{}^{AO}$. The second term involving $G_{rstu}{}^{AO}$ disappears by ZDO, since $t \neq r$ for an off-diagonal element. The third term

$$-\sum_{su,l} c_{sl}c_{ul}G_{rsut}{}^{AO} \qquad \text{becomes} \qquad -\sum_{l} c_{tl}c_{rl}G_{rtrt}{}^{AO}$$

5.6-5

since s must equal t and $u = r$ by ZDO. Also we can use the definition of bond order to further simplify this last term. We then obtain

$$F_{rt}{}^{AO} = H_{rt}{}^{AO} - \tfrac{1}{2}P_{rt}\gamma_{rt} \qquad\qquad 5.6\text{-}6$$

We use the diagonal and off-diagonal elements as defined by 5.6-4 and 5.6-6 in the usual secular equations, except that now we also neglect overlap integrals off the diagonal. The secular equations are

$$\sum_t (F_{rt} - E\delta_{rt})c_t = 0 \qquad\qquad 5.6\text{-}7$$

5.7   Configuration Interaction

Thus far we have aimed at optimizing LCAO-MO coefficients either so that single MOs would be minimized in energy or so that Slater determinant wavefunctions would have minimum energy.

Now we consider using Slater determinants as basis functions and mixing these. This will give us a linear combination of Slater determinants. Since each Slater determinant represents an electronic configuration, we are really mixing configurations. For such mixing we will use the full $\mathfrak{F}$ operator, thus including electron–electron repulsion.

As an example, let us consider the situation of a molecule (e.g., ethylene) having one electron promoted from MO k to MO l. We can write four configurations:

The Slater determinantal functions are

$$\phi_1 = |\Psi_k(1)\Psi_l(2)|, \qquad \phi_2 = |\bar{\Psi}_k(1)\bar{\Psi}_l(2)|$$

$$\phi_3 = |\Psi_k(1)\bar{\Psi}_l(2)|, \qquad \phi_4 = |\bar{\Psi}_k(1)\Psi_l(2)| \qquad 5.7\text{-}1$$

Our intention is to mix these four configurational functions in a 4×4

secular determinant

$$
\begin{array}{c|cccc}
 & \phi_1 & \phi_2 & \phi_3 & \phi_4 \\
\hline
\phi_1 & (F_{11} - E) & F_{12} & F_{13} & F_{14} \\
\\
\phi_2 & F_{21} & (F_{22} - E) & F_{23} & F_{24} \\
\\
\phi_3 & F_{31} & F_{32} & (F_{33} - E) & F_{34} \\
\\
\phi_4 & F_{41} & F_{42} & F_{43} & (F_{44} - E)
\end{array} = 0 \qquad 5.7\text{-}2
$$

where each F_{rs} matrix element is defined as

$$
F_{rs} = \int \phi_r \mathfrak{F} \phi_s \, d\tau = \int \phi_r (\mathfrak{K} + \mathfrak{G}) \phi_s \, d\tau \qquad 5.7\text{-}3
$$

where ϕ_r and ϕ_s represent Slater determinants, or permuted product functions.

We now proceed to evaluate each of the elements. In this we remember that we do not need to permute the configuration before the $\mathfrak{F}$ operator. In the case of F_{11}

$$
F_{11} = \int \Psi_k(1)\Psi_l(2)\mathfrak{F}\bar{P}\Psi_k(1)\Psi_l(2) \, d\tau_{12} = I_k + I_l + G_{klkl}{}^{\text{MO}} - G_{kllk}{}^{\text{MO}}
$$

$$5.7\text{-}4$$

This is just the energy of the closed-shell Slater determinant ϕ_1 before any mixing with other Slater determinants. We can write this result by inspection as discussed earlier or can consider the integral in detail. The first two terms arise from use of the $\mathfrak{K}$ portion of the $\mathfrak{F}$ operator, the $G_{klkl}{}^{\text{MO}}$ term comes from use of the $\mathfrak{G}$ operator without permutation, and the $-G_{kllk}{}^{\text{MO}}$ derives from use of the $\mathfrak{G}$ operator with permutation.

In exactly parallel fashion

$$
F_{22} = I_k + I_l + G_{klkl}{}^{\text{MO}} - G_{kllk}{}^{\text{MO}} \qquad 5.7\text{-}5
$$

In the case of F_{33} and F_{44} the one-electron terms and the nonpermuted two-electron term arise similarly, but spin orthogonality leads to no exchange terms. We obtain

$$
F_{33} = \int \Psi_k(1)\bar{\Psi}_l(2)\mathfrak{F}\bar{P}\Psi_k(1)\bar{\Psi}_l(2) \, d\tau = I_k + I_l + G_{klkl}{}^{\text{MO}} \qquad 5.7\text{-}6
$$

and

$$
F_{44} = I_k + I_l + G_{klkl}{}^{\text{MO}} \qquad 5.7\text{-}7
$$

Proceeding to the off-diagonal elements we find that $F_{12} = 0$. Spin orthogonality wipes out all one- and two-electron terms.

$$F_{12} = \int \Psi_k(1)\Psi_l(2)\mathfrak{F}\bar{P}\Psi_k(1)\bar{\Psi}_l(2) \, d\tau = 0 \qquad 5.7\text{-}8$$

Continuing, we have

$$F_{13} = \int \Psi_k(1)\Psi_l(2)\mathfrak{F}\bar{P}\Psi_k(1)\bar{\Psi}_l(2) \, d\tau = 0 \qquad 5.7\text{-}9$$

and similarly $F_{14} = 0$ again as a consequence of spin orthogonality. Also, F_{23} and $F_{24} = 0$ in identical fashion.

In the case of F_{34} we have

$$F_{34} = \int \Psi_k(1)\bar{\Psi}_l(2)\mathfrak{F}\bar{P}\bar{\Psi}_k(1)\Psi_l(2) \, d\tau = -G_{kllk}{}^{MO} \qquad 5.7\text{-}10$$

Here, without permutation we have spin orthogonality and only $-G_{kllk}{}^{MO}$ results. Secular determinant 5.7-2 now can be filled in as

	ϕ_1	ϕ_2	ϕ_3	ϕ_4
ϕ_1	$(I_k + I_l + G_{klkl}{}^{MO}$ $- G_{kllk}{}^{MO} - E)$	0	0	0
ϕ_2	0	$(I_k + I_l + G_{klkl}{}^{MO}$ $- G_{kllk}{}^{MO} - E)$	0	0
ϕ_3	0	0	$(I_k + I_l + G_{klkl}{}^{MO}$ $- E)$	$-G_{kllk}{}^{MO}$
ϕ_4	0	0	$-G_{kllk}{}^{MO}$	$(I_k + I_l + G_{klkl}$ $- E)$

$$5.7\text{-}11$$

Thus ϕ_1 and ϕ_2 are final eigenfunctions. The 2×2 involving ϕ_3 and ϕ_4 has equal diagonal elements and is reminiscent of the Hückel ethylene problem. Thus this can be diagonalized by addition–subtraction, thus giving $(1/\sqrt{2})(\phi_3 + \phi_4)$ and $(-1/\sqrt{2})(\phi_3 - \phi_4)$ as eigenfunctions. The former gives the lower energy eigenvalue, i.e.,

$$E = I_k + I_l + G_{klkl}{}^{MO} - G_{kllk}{}^{MO} \qquad \text{for} \quad (1/\sqrt{2})(\phi_3 + \phi_4) \quad 5.7\text{-}12$$

The latter gives an energy of

$$E = I_k + I_l + G_{klkl}{}^{MO} + G_{kllk}{}^{MO} \qquad 5.7\text{-}13$$

Hence the original problem of configuration interaction in this case has led to the three degenerate wavefunctions

$$\phi_1, \quad \phi_2 \quad \text{and} \quad (1/\sqrt{2})(\phi_3 + \phi_4) \qquad\qquad 5.7\text{-}14$$

all with an energy $I_k + I_l + G_{klkl}{}^{MO} - G_{kllk}{}^{MO}$ and also to a higher energy eigenfunction

$$(1/\sqrt{2})(\phi_3 - \phi_4) \qquad \text{with an energy} \quad I_k + I_l + G_{klkl}{}^{MO} + G_{kllk}{}^{MO} \quad 5.7\text{-}15$$

The first three (Eq. 5.7-14) constitute the three components of a triplet species and the last one (Eq. 5.7-15) is an excited singlet. The energy difference between these is twice the exchange integral, i.e.,

$$\Delta E = 2G_{kllk}{}^{MO} \qquad\qquad 5.7\text{-}16$$

and this proves to be general.

This result gives us the general form in which to write excited singlets and triplets. These have the form

$$
{}^{1}_{3}\Phi = (1/\sqrt{2})\{|\Psi_1(1)\bar{\Psi}_1(2)\Psi_2(3)\bar{\Psi}_2(4)\cdots\Psi_l(2n-1)\bar{\Psi}_m(2n)|
$$
$$
\mp \; |\Psi_1(1)\bar{\Psi}_1(2)\Psi_2(3)\bar{\Psi}_2(4)\cdots\bar{\Psi}_l(2n-1)\Psi_m(2n)|\} \qquad 5.7\text{-}17
$$

where the singlet $^1\Phi$ has the minus sign and the triplet $^3\Phi$ is assigned the plus sign.

Another example of configuration interaction arises in connection with the use of correlation diagrams for organic reactions. A forbidden ground-state reaction is one in which an occupied bonding MO becomes antibonding during the reaction. Figure 5.7-A shows the situation where a closed-shell reactant configuration with only bonding MOs occupied adiabatically transforms itself into a product set of MOs. Since reactant

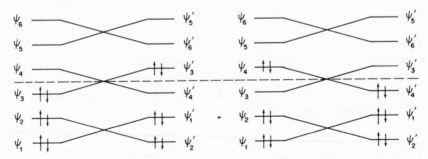

FIG. 5.7-A. Ground-state reactant giving doubly excited state product.

FIG. 5.7-B. Doubly excited state reactant giving ground-state product.

MO 3 becomes antibonding during the reaction, the product configuration is doubly excited. The entire wavefunction for this configuration can be expressed as Slater determinant Φ_I. It is clear that Φ_I becomes increasingly high in energy as the reaction proceeds:

$$\Phi_I = |\Psi_1(1)\,\bar{\Psi}_1(2)\,\Psi_2(3)\,\bar{\Psi}_2(4)\,\Psi_3(5)\,\bar{\Psi}_3(6)| \qquad 5.7\text{-}18$$

Although the present situation is written for a six-electron, six-MO species, the example applies generally.

A second configuration is Φ_{II} given in Fig. 5.7-B. This reaction diagram starts with a doubly excited configuration and ends with a ground-state one. Clearly, this begins as being of rather high energy and ends being low in energy. The configurational wavefunction is given by

$$\Phi_{II} = |\Psi_1(1)\,\bar{\Psi}_1(2)\,\Psi_2(3)\,\bar{\Psi}_2(4)\,\Psi_4(5)\,\bar{\Psi}_4(6)| \qquad 5.7\text{-}19$$

We now consider the consequences of interaction of these two Slater determinantal wavefunctions (i.e., configuration interaction). The secular matrix is given by

$$\begin{array}{cc} & \begin{matrix} \Phi_I & \quad \Phi_{II} \end{matrix} \\ \begin{matrix} \Phi_I \\ \\ \Phi_{II} \end{matrix} & \begin{bmatrix} F_{I\,I} & F_{I\,II} \\ \\ F_{II\,I} & F_{II\,II} \end{bmatrix} \end{array} \qquad 5.7\text{-}20$$

where

$$F_{uv} = \int \Phi_u \mathcal{F} \Phi_v \, d\tau \qquad 5.7\text{-}21$$

We obtain $F_{I\,I}$ and $F_{II\,II}$ in the usual way as

$$F_{I\,I} = 2I_1 + 2I_2 + 2I_3 + G_{1111} + G_{2222} + G_{3333} + 4G_{1212} + 4G_{1313}$$
$$+ 4G_{2323} - 2G_{1221} - 2G_{1331} - 2G_{2332} \qquad 5.7\text{-}22a$$

$$F_{II\,II} = 2I_1 + 2I_2 + 2I_4 + G_{1111} + G_{2222} + G_{4444} + 4G_{1212} + 4G_{1414}$$
$$+ 4G_{2424} - 2G_{1221} - 2G_{1441} - 2G_{2442} \qquad 5.7\text{-}22b$$

In the case of $F_{I\,II}$ we evaluate the integral

$$F_{I\,II} = \int \Psi_1(1)\,\bar{\Psi}_1(2)\,\Psi_2(3)\,\bar{\Psi}_2(4)\,\Psi_3(5)\,\bar{\Psi}_3(6)$$
$$\times \mathcal{F}\bar{P}\Psi_1(1)\,\bar{\Psi}_1(2)\,\Psi_2(3)\,\bar{\Psi}_2(4)\,\Psi_4(5)\,\bar{\Psi}_4(6)\, d\tau$$
$$= G_{3344}{}^{MO} = [34\,|\,34] \qquad 5.7\text{-}23$$

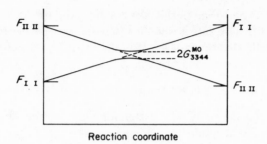

Reaction coordinate

FIG. 5.7-C.   Noncrossing of ϕ_I and ϕ_{II} along the reaction coordinate as a result of configuration interaction.

We evaluate $G_{3344}{}^{MO} = [34 \mid 34]$ as follows:

$$[34 \mid 34] = \sum_{r,s} \omega_{r,34}\gamma_{rs}\omega_{s,34} = \tilde{\omega}_{34}\Gamma\omega_{34} \qquad 5.7\text{-}24$$

and find that in general this quantity does not vanish. With a nonvanishing off-diagonal element $F_{I\ II}$ we find that the secular matrix in expression 5.7-20 will diagonalize to give energies higher and lower than $F_{I\ I}$ and $F_{II\ II}$ even when these two approach one another and become degenerate. Thus, there is a splitting of states which results in noncrossing as shown in Fig. 5.7-C.

5.8   General Expressions for Use in Configuration Interaction

Configuration interaction is normally used for more general purposes than in the two examples presented thus far. Usually, after one has obtained one-electron MOs, either of the Hückel or SCF variety, a further improvement in the energy of the ground state can be obtained by admixing the closed-shell ground state with as large a number of excited configurations as is practical. Also, if possible, doubly excited configurations should be included. From such admixing one obtains the energies of both singlets and triplets. The wavefunctions obtained are in the form of linear combinations of Slater determinantal polyelectron wavefunctions. While the eigenfunctions thus are not quite as convenient to work with as simple one-electron MOs, as in Hückel or SCF approximations, they nevertheless can be used to give all the desired physical properties as electron densities, bond orders, and so on. Furthermore, the energies for the lower excited states are found to be decreased relative to the unmixed Slater determinantal forms, and thus configuration interaction also gives us better excited state energies and good excited state wavefunctions.

In order to carry out configuration interaction practically, we need to use general expressions for the matrix elements in order to avoid having to calculate these from first principles each time we do such a calculation. Since the expressions given below can be derived by the reader using the methods given thus far, the details are omitted and it is suggested that the reader try deriving some of these as problems. For matrix elements between singly excited configurations:*

$$
{}^{1}F_{km}{}^{ln} = \int {}^{1}\Phi_{k}{}^{l} \, \mathfrak{F} \, {}^{1}\Phi_{m}{}^{n} \, d\tau = \delta_{km}\delta_{ln}E_{0} + \delta_{km}F_{ln} - \delta_{ln}F_{km} + {}^{1}G \qquad 5.8\text{-}1
$$

where the superscript ${}^{3}_{1}$ refers to the two possible singly excited states (i.e., singlet and triplet) and one selects either the 1 or the 3 superscript. The k, l, m, and n refer to four MOs of the Hückel or SCF variety and the δ_{pq}'s are the usual Kronecker deltas, which are zero if $p \neq q$ and unity if $p = q$. E_{0} is the energy of the closed-shell ground-state configuration (i.e., in absence of configuration interaction). The F terms are defined as

$$
F_{ln} = H_{ln}{}^{\text{MO}} + 2 \sum_{w}^{N} G_{wlwn}{}^{\text{MO}} - \sum_{w}^{N} G_{wlnw}{}^{\text{MO}} \qquad 5.8\text{-}2a
$$

$$
F_{km} = H_{km}{}^{\text{MO}} + 2 \sum_{w}^{N} G_{wkwm}{}^{\text{MO}} - \sum_{w}^{N} G_{wkmw}{}^{\text{MO}} \qquad 5.8\text{-}2b
$$

$$
{}^{1}G = G_{knlm} - G_{knml} \mp G_{knlm} \qquad 5.8\text{-}3
$$

The summations are over all N MOs of the ground configuration. In Eq. 5.8-3 the plus or minus sign depends on whether the singlet or the triplet is being considered.

Matrix elements between singlet and triplet configurations disappear. Interaction elements between the ground-state configuration and singly excited configurations are given by

$$
{}^{1}F_{0k}{}^{l} = \sqrt{2}F_{kl} = \sqrt{2}(H_{kl}{}^{\text{MO}} + 2 \sum_{w} G_{kwlw}{}^{\text{MO}} - \sum_{w} G_{kwwl}{}^{\text{MO}}) \qquad 5.8\text{-}4
$$

In the case where the MOs used are SCF orbitals, then Brillouin's theorem leads to zero matrix interaction elements between the closed-shell ground state and singly excited singlet configurations (i.e., as in Eq. 5.8-4). In any case matrix elements between different multiplicity configurations also vanish.

* The notation $\Phi_{k}{}^{l}$ is used to represent the configuration in which an electron has been promoted from MO k to MO l. The configuration is given in Slater determinantal form (e.g., note Eq. 5.7-17).

One interesting point arises if we attempt to take a closed-shell Slater determinant consisting of non-SCF MOs and admix each of these with small increments of other members of the complete set of MOs. Let us take a four-MO system (e.g., butadiene) as an example. We write the usual Slater determinant, except that to each space-spin MO we add an increment of all of the other MOs:

$$\Phi = |\{\Psi_1(1) + \delta_{12}\Psi_2(1) + \delta_{13}\Psi_3(1) + \delta_{14}\Psi_4(1)\}$$
$$\times \{\bar{\Psi}_1(2) + \delta_{12}\bar{\Psi}_2(2) + \delta_{13}\bar{\Psi}_3(2) + \delta_{14}\bar{\Psi}_4(2)\}$$
$$\times \{\Psi_2(3) + \delta_{21}\Psi_1(3) + \delta_{23}\Psi_3(3) + \delta_{24}\Psi_4(3)\}$$
$$\times \{\bar{\Psi}_2(4) + \delta_{21}\bar{\Psi}_1(4) + \delta_{23}\bar{\Psi}_3(4) + \delta_{24}\bar{\Psi}_4(4)\}| \qquad 5.8\text{-}5$$

If we expand this Slater determinant, we will obtain a total of 4^4 determinants. We obtain

$$\Phi = |\Psi_1(1)\bar{\Psi}_1(2)\Psi_2(3)\bar{\Psi}_2(4)| + \qquad\qquad\qquad\qquad\qquad\text{a}$$

$$\delta_{12}\{|\Psi_2(1)\bar{\Psi}_1(2)\Psi_2(3)\bar{\Psi}_2(4)| + |\Psi_1(1)\bar{\Psi}_2(2)\Psi_2(3)\bar{\Psi}_2(4)|\} + \quad\text{b}$$

$$\delta_{21}\{|\Psi_1(1)\bar{\Psi}_1(2)\Psi_1(3)\bar{\Psi}_2(4)| + |\Psi_1(1)\bar{\Psi}_1(2)\Psi_2(3)\bar{\Psi}_1(4)|\} + \quad\text{c}$$

$$\delta_{13}\{|\Psi_3(1)\bar{\Psi}_1(2)\Psi_2(3)\bar{\Psi}_2(4)| + |\Psi_1(1)\bar{\Psi}_3(2)\Psi_2(3)\bar{\Psi}_2(4)|\} + \quad\text{d}$$

$$\delta_{14}\{|\Psi_4(1)\bar{\Psi}_1(2)\Psi_2(3)\bar{\Psi}_2(4)| + |\Psi_1(1)\bar{\Psi}_4(2)\Psi_2(3)\bar{\Psi}_2(4)|\} + \quad\text{e}$$

$$\delta_{23}\{|\Psi_1(1)\bar{\Psi}_1(2)\Psi_3(3)\bar{\Psi}_2(4)| + |\Psi_1(1)\bar{\Psi}_1(2)\Psi_2(3)\bar{\Psi}_3(4)|\} + \quad\text{f}$$

$$\delta_{24}\{|\Psi_1(1)\bar{\Psi}_1(2)\Psi_4(3)\bar{\Psi}_2(4)| + |\Psi_1(1)\bar{\Psi}_1(2)\Psi_2(3)\bar{\Psi}_4(4)|\} \qquad\text{g}$$

plus terms representing doubly excited configurations and weighted with δ_{13}^2, δ_{14}^2, δ_{23}^2, δ_{24}^2, $\delta_{13}\delta_{14}$, etc.   5.8-6

For MOs which are nearly SCF the improvements to each MO (i.e., the δ's) will be small and we can neglect the squared terms, the terms consisting of the product of two different δ's, and terms of still higher order in the δ's.

If we now turn our attention to the expansion in Eq. 5.8-6, we note that the term on line a is just the original closed-shell Slater determinant before improvement. Interestingly, the determinants on lines b and c are zero since these are determinants each of which has two identical columns, and such determinants vanish. The Slater determinantal wavefunctions in lines b and c are seen to result from admixture (note the coefficients δ_{12} and δ_{21}) of MOs within the closed shell with other closed-shell MOs which need improvement. The result is that there is no improvement by different linear combinations of the same MOs already utilized.

Even more interesting are the terms in lines d, e, f, and g. Each of these can be seen to be a Slater determinantal combination corresponding to a singlet excited state, although one must note that the form is slightly different than that given in Eq. 5.7-17 for singlets. One can permute the first two terms of the second determinant in lines d and e and the last two terms of the first Slater determinant in each of lines f and g. In doing this we obtain the negative sign between the two determinants and thus have the characteristic excited singlet form.

This means then that improvement of the original non-SCF single determinantal wavefunction to self-consistency is equivalent to configuration interaction mixing in all singly excited state configurations. Also, we have here a proof of Brillouin's theorem, since as the variation needed to reach self-consistency approaches zero, the contribution of the singly excited configurations also becomes zero. Thus for SCF MOs, there will be no matrix interaction elements between the ground-state Slater determinant wavefunction and the singly excited singlet configurational wavefunctions. Actually, one might use Eqs. 5.8-5 and 5.8-6 to obtain an approximate SCF set of MOs by using the coefficients weighting the singly excited singlet determinantal wavefunctions as derived from a simple configuration interaction calculation. This would give only the bonding MOs. However, if one wanted the antibonding MOs, one could use a similar treatment based on an initial determinant in which other sets of MOs (e.g., all antibonding) were doubly occupied.

Still another point of interest concerns the significance of the matrix elements (e.g., F_{ln}) in Eqs. 5.8-2a and 5.8-2b. We take Eq. 5.8-2a and substitute for $H_{ln}{}^{\mathrm{MO}}$, for $G_{wlwn}{}^{\mathrm{MO}}$ and for $G_{wlnw}{}^{\mathrm{MO}}$ as follows:

$$H_{ln}{}^{\mathrm{MO}} = \sum_{r,t} c_{rl}c_{tn}H_{rt}{}^{\mathrm{AO}} \qquad\qquad 5.8\text{-}7$$

$$G_{wlwn}{}^{\mathrm{MO}} = G_{lwnw}{}^{\mathrm{MO}} = \sum_{rstu} c_{rl}c_{sw}c_{tn}c_{uw}G_{rstu}{}^{\mathrm{AO}} \qquad (\text{note Eq. 5.3-1}) \qquad 5.8\text{-}8a$$

$$G_{wlnw}{}^{\mathrm{MO}} = G_{lwwn}{}^{\mathrm{MO}} = \sum_{rstu} c_{rl}c_{sw}c_{uw}c_{tn}G_{rsut}{}^{\mathrm{AO}} \qquad (\text{again note Eq. 5.3-1})$$

$$5.8\text{-}8b$$

This affords

$$F_{ln}{}^{\mathrm{MO}} = \sum_{rt} c_{rl}c_{tn}\{H_{rt}{}^{\mathrm{AO}} + \sum_{su,w} c_{sw}c_{uw}(2G_{rstu}{}^{\mathrm{AO}} - G_{rsut}{}^{\mathrm{AO}})\} \qquad 5.8\text{-}9$$

However, we notice that the quantity in the braces has the same form as $F_{rt}{}^{\mathrm{AO}}$ of Eq. 5.5-2 and substitution into 5.8-9 gives us

$$F_{ln}{}^{\mathrm{MO}} = \sum_{rt} c_{rl}c_{tn}F_{rt}{}^{\mathrm{AO}} \qquad\qquad 5.8\text{-}10$$

Thus we see that $F_{ln}{}^{\mathrm{MO}}$ is just a matrix element between MOs l and n quite as $H_{ln}{}^{\mathrm{MO}}$ would be except that here the full $\mathfrak{F}$ operator is used rather than just $\mathfrak{K}$. If ψ_l and ψ_n are eigenfunctions of $\mathfrak{F}$, then off-diagonal elements $F_{ln}{}^{\mathrm{MO}}$ will vanish unless $l = n$ (i.e., for F_{11}). If l does equal n, note that Eq. 5.8-2a becomes identical with 5.2-25 which gives the single MO energy including electron–electron repulsion.

5.9 Three-Dimensional Hückel Theory; The Extended Hückel Treatment

The methods we have used in Hückel treatments thus far have assumed either a set of p orbitals or some other truncated set of orbitals which has excluded part of the molecular framework. The part excluded often has been a planar σ framework or in other instances it has been a part of the molecule not particularly of interest and assumed not to interact with the truncated, delocalized system of orbitals considered. While symmetry does allow one to separate a planar σ framework from the π system of p orbitals perpendicular to this plane, the dissection into two separate sets of MOs, one studied and the other disregarded, is not ideal.

If we decide to include all valence shell atomic orbitals in a calculation and thus admix three p orbitals and one carbon $2s$ for each carbon atom and admix a hydrogen $1s$ for each hydrogen atom, we have what is termed extended Hückel theory. Alternatively, we could use a basis set consisting of hybrid orbitals but then orbitals at any carbon have H_{rs} matrix elements between them while p_x, p_y, p_z, and $2s$ orbitals centered at each carbon differ in symmetry and do not have such matrix elements.

If we select the p_x, p_y, p_z, and $2s$ orbitals at each center as our basis, it

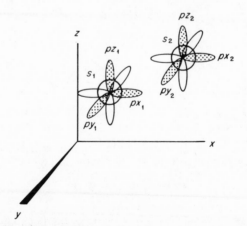

Fig. 5.9-A. Orientation of basis orbitals at two carbon atoms of a molecule.

is clear that the Hückel solution requires knowledge of the overlap between orbitals at different carbons and oriented at odd angles relative to one another (e.g., note Fig. 5.9-A). In Fig. 5.9-A we have depicted only two of the many atoms of a typical system of interest; but the relations derived will apply to all sets of two adjacent atomic centers.

Thus, using the methods described previously for obtaining overlap between orbitals with unusual geometric relationships relative to one another, we can derive the expressions in the following equations:

$$S_{xx} = (\Delta x/\rho)^2 S_{\sigma\sigma} + [(\Delta y^2 + \Delta z^2)/\rho^2] S_{\pi\pi} \qquad \text{5.9-1}$$

$$S_{yy} = (\Delta y/\rho)^2 S_{\sigma\sigma} + [(\Delta x^2 + \Delta z^2)/\rho^2] S_{\pi\pi} \qquad \text{5.9-2}$$

$$S_{zz} = (\Delta z/\rho)^2 S_{\sigma\sigma} + [(\Delta x^2 + \Delta y^2)/\rho^2] S_{\pi\pi} \qquad \text{5.9-3}$$

$$S_{xy} = (\Delta y \Delta x/\rho^2) S_{\sigma\sigma} - (\Delta x \Delta y/\rho^2) S_{\pi\pi} \qquad \text{5.9-4}$$

$$S_{xz} = (\Delta x \Delta z/\rho^2) S_{\sigma\sigma} - (\Delta x \Delta z/\rho^2) S_{\pi\pi} \qquad \text{5.9-5}$$

$$S_{yz} = (\Delta y \Delta z/\rho^2) S_{\sigma\sigma} - (\Delta y \Delta z/\rho^2) S_{\pi\pi} \qquad \text{5.9-6}$$

$$S_{xs} = (\Delta x/\rho) S_{\sigma s} \qquad \text{5.9-7}$$

$$S_{ys} = (\Delta y/\rho) S_{\sigma s} \qquad \text{5.9-8}$$

$$S_{zs} = (\Delta z/\rho) S_{\sigma s} \qquad \text{5.9-9}$$

Here S_{xx}, S_{yy}, S_{zz}, S_{xy}, S_{xz}, and S_{yz} give us the overlap integrals between two p orbitals with the indicated orientations and at two centers. Similarly, S_{xs}, S_{ys}, and S_{zs} give us the overlap integrals between a p_x, p_y, or p_z orbital and a carbon $2s$ orbital at another center. The ρ's are the distances between the two atomic centers, $S_{\sigma\sigma}$ is a standard overlap integral, available from tables, for two p orbitals oriented colinearly at the distance ρ apart. Similarly, $S_{\pi\pi}$ is the standard overlap integral value, again available from tables, where the two p orbitals are oriented parallel to one another. $S_{\sigma s}$ is the standard integral between an s orbital and a p orbital at the second center aimed at the s orbital. The Δx, Δy, and Δz values are just the difference in x, y, and z coordinates at the two centers considered. Finally, in using these relationships, we adopt the convention that $S_{\sigma s}$ is positive if the first subscript is σ and the second is s while this overlap is negative if the first subscript is s and the second is σ, as in $S_{s\sigma}$. $S_{\sigma\sigma}$ is always negative.

These relationships thus give us the required overlaps between the basis orbitals, although in terms of standard overlap integrals.

With the overlap integrals available, we now need to discuss the H_{rs} elements, namely the resonance integrals. For the diagonal elements (i.e., the H_{rr}'s) the valence state ionization potential of the given element for the

s orbital or p orbital used is generally used directly. Thus, for hydrogen $1s$ H_{rr} is taken as -13.60 eV (i.e., the valence state ionization potential). For carbon $2s$ we use -21.01 eV, and for carbon $2p$ we use -11.27 eV.

For the off-diagonal matrix elements (H_{rs} where $r \neq s$) the Mulliken "magic formula" in Eq. 5.9-10 is used:

$$H_{rs} = K(H_{rr} + H_{ss})S_{rs}/2 \qquad\qquad 5.9\text{-}10$$

This provides a resonance integral which is proportional to the overlap integral and which gives an energy that is intermediate between the valence ionization potentials of the two orbitals r and s. K is a proportionality constant commonly taken about 1.75.

With the $[\mathbf{H} - E\mathbf{S}]$ matrix elements available, it remains to diagonalize this. However, we notice that the eigenvalue symbol E occurs in both diagonal and off-diagonal terms. One approach to solving the diagonalization problem is first to diagonalize the $\mathbf{S}$ matrix, as in Eq. 5.9-11, to give the $\mathbf{S}_d$ matrix. The Jacobi process described earlier can be used or some alternative, less time-consuming method (e.g., Gram–Schmidt orthonormalization) can be employed. Thus, there is a $\mathbf{T}$ matrix which can be used in a similarity transformation to convert $\mathbf{S}$ to $\mathbf{S}_d$. Further, we could premultiply $\mathbf{S}_d$ by a matrix $\mathbf{S}_d^{-1}$ in which each diagonal element is the reciprocal of the diagonal elements of the $\mathbf{S}_d$ matrix; this would give us the unit matrix (i.e., $\mathbf{1}$) with all ones along the diagonal and zeros elsewhere. Alternatively, we can accomplish the same result by employing the $\mathbf{S}_d^{-1/2}$, or $\mathbf{V}$, matrix twice, once before and once after as in a similarity transformation. We can do this since both $\mathbf{V}$ and $\mathbf{S}_d$ are diagonal matrices and thus commute (the order can be reversed):

$$\mathbf{\tilde{T}ST} = \mathbf{S}_d \qquad\qquad 5.9\text{-}11$$

and

$$\mathbf{\tilde{V}\tilde{T}STV} = \mathbf{\tilde{V}S}_d\mathbf{V} = \mathbf{1} \qquad\qquad 5.9\text{-}12\text{a}$$

If we define the product $\mathbf{TV}$ as $\mathbf{U}$, then 5.9-12a can be more simply rewritten as

$$\mathbf{\tilde{U}SU} = \mathbf{1} \qquad\qquad 5.9\text{-}12\text{b}$$

We can now apply this similarity transformation to the entire secular matrix and effectively remove the E terms from the off-diagonal elements as a consequence of the S_{rs} terms disappearing from off the diagonal. Thus we can see in Eq. 5.9-13 that applying the similarity transformation using the $\mathbf{U}$ matrix converts the secular matrix into a form lacking off-diagonal E terms:

$$\mathbf{\tilde{U}}[\mathbf{H} - E\mathbf{S}]\mathbf{U} = \mathbf{\tilde{U}HU} - E\mathbf{\tilde{U}SU} = [\mathbf{H}' - E\mathbf{1}] \qquad\qquad 5.9\text{-}13\text{a}$$

where

$$\mathbf{H'} = \mathbf{\tilde{U}HU} \qquad\qquad 5.9\text{-}13b$$

Hence solving the original secular matrix, including overlap, for its eigenvalues is equivalent to diagonalizing the new $\mathbf{H'}$ matrix which is expressed in terms of a new basis where the orbitals are orthogonal. To effect this diagonalization we proceed as usual:

$$\mathbf{\tilde{C}'}[\mathbf{H'} - E\mathbf{1}]\mathbf{C'} = 0 \qquad\qquad 5.9\text{-}14a$$

or

$$\mathbf{\tilde{C}'\tilde{U}}[\mathbf{H} - E\mathbf{S}]\mathbf{UC'} = \mathbf{\tilde{C}}[\mathbf{H} - E\mathbf{S}]\mathbf{C} = 0 \qquad\qquad 5.9\text{-}14b$$

Here Eq. 5.9-14b formulates the diagonalization problem in the old basis where there is overlap between nonorthogonal orbitals while Eq. 5.9-14a formulates the diagonalization problem in terms of the new, orthogonalized basis.

If we proceed as in Eq. 5.9-14a, we note that in addition to obtaining the eigenvalues, which are independent of which basis we select, we obtain the eigenvector matrix $\mathbf{C'}$. However, this set of eigenvectors (or eigenfunctions) is not particularly useful since it gives weightings of the orthogonalized basis orbitals rather than the set with which we began. However, reference to Eq. 5.9-14b reveals that we can easily convert this to the desired $\mathbf{C}$ matrix:

$$\mathbf{C} = \mathbf{UC'} \qquad\qquad 5.9\text{-}15$$

This relationship is formulated in Eq. 5.9-15.

The extended Hückel method thus can be carried out on organic systems including all bonds, both σ and π, and in fact it does not differentiate between these. As in the two-dimensional Hückel treatment it does not take into account electron–electron repulsion and exchange. In practice it gives useful results when used qualitatively but energy differences predicted are usually much too large.

5.10   Three-Dimensional SCF Methods

Just as it was possible to improve two-dimensional Hückel theory by proceeding to two-dimensional SCF approaches, three-dimensional quantum mechanical treatments are improved by proceeding to methods including electron–electron interaction.

We begin with the Pople SCF equations 5.6-4 for diagonal matrix elements F_{rr} and 5.6-6 for off-diagonal elements F_{rt}. We first concentrate attention on the H_{rr} and H_{rt} terms in these expressions. For this we use the one-electron Hamiltonian operator $\mathcal{H}$:

$$\mathcal{H} = -(h^2/8m\pi^2)\,\nabla^2 - V_A - V_B - V_C - \cdots \qquad 5.10\text{-}1$$

where ∇^2 represents the sum of second partial derivative operators with respect to the three coordinates. More commonly this is written in units such that the coefficient of ∇^2 is $-\frac{1}{2}$, and for simplicity this will be used henceforth. Accordingly, H_{rr} may be written as

$$H_{rr} = \int \chi_r(1)\left[-\tfrac{1}{2}\nabla^2 - V_A - V_B - V_C - \cdots\right]\chi_r(1)\,d\tau \qquad 5.10\text{-}2\text{a}$$

or

$$(r|-\tfrac{1}{2}\nabla^2 - V_A - V_B - V_C - \cdots|r) \qquad 5.10\text{-}2\text{b}$$

The notation in Eq. 5.10-2b is just an alternative way of writing integrals. More important, we note that we can partition the integral into two separate integrals. One gives us the energy of an electron in AO r. We designate this U_{rr} and note that it includes both the kinetic energy of the electron in this AO on atom A and also the potential energy of attraction for the nucleus, or core, of this atom:

$$H_{rr} = (r|-\tfrac{1}{2}\nabla^2 - V_A|r) - \sum_{B \neq A} (r|V_B|r) \qquad 5.10\text{-}3\text{a}$$

$$= U_{rr} - \sum_{B \neq A} (r|V_B|r) \qquad 5.10\text{-}3\text{b}$$

The other term derived from the original integral consists of the sum of $(r|V_B|r)$ integrals and these represent the potential energy of attraction for the electron in AO r for all the nuclei other than A (i.e., where B corresponds to each atom).

Turning attention to the off-diagonal terms H_{rt} we can again use the same Hamiltonian operator in Eq. 5.10-1 and obtain

$$H_{rt} = (r|-\tfrac{1}{2}\nabla^2 - V_A - V_B - V_C - \cdots|t) \qquad 5.10\text{-}4$$

If atomic orbitals r and t are on the same atom A, then the portion of the integral corresponding to use of the operators $-\frac{1}{2}\nabla^2$ and $-V_A$ vanishes due to symmetry; that is, r and t necessarily differ in symmetry (e.g., one being p_y and the other s, etc.) while the operator is totally symmetric. Also with r and t on the same atom but being different orbitals, the portions of the integral deriving from use of the other operators (e.g., V_B) are zero due to zero differential overlap.

Where atomic orbitals r and t are on different atoms, A and B, it is convenient to partition the integral in Eq. 5.10-4 into a part including

$(-\nabla^2 - V_A - V_B)$ and a second part using $(-V_C - V_D - \cdots)$. Thus,

$$H_{rt} = (r|-\tfrac{1}{2}\nabla^2 - V_A - V_B|t) - \sum_{C \neq A,B} (r|V_C|t) \qquad 5.10\text{-}5$$

The first operator affords the kinetic and potential energy contributions of an electron spread between orbitals r and t on atoms A and B; this includes the attraction for nuclei A and B. Thus in Eq. 5.10-5, the first term is just the usual resonance integral β_{rt} between the two orbitals r and t. However, the second term includes only the potential energy contributions due to attraction of the electron spread between orbitals r and t by nuclei C, D, etc. These are taken as small and neglected in CNDO calculations. In summary H_{rt} is given by

$$H_{rt} = \beta_{rt} = S_{rt}\beta_{rt}^0 \qquad 5.10\text{-}6$$

We should note that the β_{rt}^0, in contrast to the usual resonance integrals connoted by this symbol, is independent of the extent of overlap between AOs r and t and is just a function of the energies of the orbitals r and t.

We now substitute the value of H_{rr} (note Eq. 5.10-3b) into the Pople SCF expression for F_{rr} (Eq. 5.6-4) and also remove the restriction over the summation so that s may equal r; to compensate for the latter we subtract $q_r\gamma_{rr}$.* This gives us

$$F_{rr} = U_{rr} - \sum_{B \neq A} (r|V_B|r) - \tfrac{1}{2}q_r\gamma_{rr} + \sum_{s,r;s \text{ on } A} q_s\gamma_{rs} + \sum_{s,r;s \text{ on } B} q_s\gamma_{rs} \qquad 5.10\text{-}7$$

Here we have also dissected the summation of $q_s\gamma_{ss}$ terms into those which correspond to orbitals on atom A and those on atom B, remembering that atomic orbital r is on A.

Now, in CNDO (complete neglect of differential overlap) treatments the approximation is made that all repulsion integrals are a function only of which two atoms the two orbitals are on and are independent of the orientation and hybridization of each pair of AOs involved in a repulsion integral. Thus, in Eq. 5.10-7 we label γ_{rr} as γ_{AA}. In the fourth term in Eq. 5.10-7 we set all the γ_{rs} terms equal to the same γ_{AA} since in this summation both orbitals r and s are on atom A. Clearly this assumption is a bit drastic since the integrals really differ appreciably; for example, γ_{rr} is much larger than γ_{rs}, for in the former case two electrons repelling one another are confined in one AO while in the latter they are in different AOs albeit on the same atom. Finally, for the last term, the γ_{rs} terms are set equal to a common parameter γ_{AB}. Here the error introduced tends to be smaller.

---

* Here for simplicity we omit the prime on q_r used earlier to signify total electron density on atom r. We recognize this term by its single subscript not involving any MO.

With these substitutions, Eq. 5.10-7 becomes

$$F_{rr} = U_{rr} + (q_A - \tfrac{1}{2}q_r)\gamma_{AA} + \sum_{B \neq A} (q_B\gamma_{AB} - V_{rB}) \qquad 5.10\text{-}8$$

The term V_{rB} represents the attraction for an electron in AO r on A for the nucleus of atom B; this is the same term as the last one in Eq. 5.10-3.

We obtain the off-diagonal term from Eq. 5.6-6 by substitution for H_{rt} and use of a generalized repulsion integral:

$$F_{rt} = S_{rt}\beta^0 - \tfrac{1}{2}P_{rt}\gamma_{AB} \qquad 5.10\text{-}9$$

Finally, there is an advantage to rewriting the diagonal terms with inclusion of the definition for total electron density on atom B as

$$q_B = Z_B - Q_B \qquad 5.10\text{-}10$$

where Q_B is the charge on atom B and Z_B is the nuclear charge of this atom. We obtain

$$F_{rr} = U_{rr} + (q_A - \tfrac{1}{2}q_r)\gamma_{AA} + \sum_{B \neq A} [-Q_B\gamma_{AB} + (Z_B\gamma_{AB} - V_{rB})] \quad 5.10\text{-}11$$

The quantity $(Z_B\gamma_{AB} - V_{rB})$ is termed a penetration integral. It can be seen to represent the energy of an electron in atomic orbital r or atom A as determined by its attraction by the positive core of atom B (i.e., a negative or stabilizing energy effect) and its repulsion by the full complement of electrons (Z_B in number) surrounding atom B.

In the CNDO/2 treatment, which is representative of a number of three-dimensional SCF approaches, the following assumptions are made in approximating the diagonal and off-diagonal matrix elements of Eqs. 5.10-9 and 5.10-11.

(a)   First, as in many of the methods, overlap is neglected in the secular determinant. Thus all of the coefficients are normalized so that the sums of the squares in any eigenfunction add up to unity. Also, as the name "complete neglect of differential overlap" implies, differential overlap is neglected so that only repulsion integrals of the type γ_{rs} are retained. This has already been assumed in these equations. Furthermore, these integrals are given the value γ_{AB} which is assumed to be independent of the type and orientation of the two orbitals r and s and dependent only on the distance between the two atoms bearing r and s (i.e., atoms A and B). Despite neglect of differential overlap for most integrals, the approximation is not invoked for the resonance integrals; as in Hückel theory, this would prove too drastic.

(b)   Resonance integrals are taken as proportional to overlap integrals despite the latter being neglected subsequently. This is again parallel to Hückel theory.

(c) The penetration integrals are neglected. This means that we are assuming, as an approximation, that $V_{rB} = Z_B \gamma_{AB}$. This is equivalent to saying that the attraction of an electron by a nucleus is exactly counter-balanced by the repulsion of the electron by the valence shell electrons of that atom.

(d) Finally, the term U_{rr} is derived from Eq. 5.10-12 which uses both the ionization potential of the orbital of interest

$$U_{rr} = -\tfrac{1}{2}(I_r + A_r) - (Z_r - \tfrac{1}{2})\gamma_{rr} \qquad 5.10\text{-}12$$

as well as its electron affinity (A_r).

In using such three-dimensional self-consistent field calculations, one often can obtain satisfying ground-state properties which are in good agreement with experiment. Ideally, one would then like to use the wave-functions for configuration interaction so that not only ground state but also excited states might be better approximated. One difficulty is that what is good parametrization for ground-state SCF methods without con-figuration interaction is not optimum for inclusion of configuration inter-tion. Also, configuration interaction with so many configurations possible proves less than simple.

Problems

1. Which of the following integrals is zero? Show why.

(a) $\displaystyle\int \Psi_1(1)\,\bar\Psi_1(2)\,\Psi_2(3)\,\bar\Psi_2(4)\,\mathfrak{F}\Psi_1(1)\,\bar\Psi_1(2)\,\bar\Psi_2(3)\,\Psi_2(4)\;d\tau.$

(b) $\displaystyle\int \Psi_1(1)\,\bar\Psi_1(2)\,\Psi_2(3)\,\bar\Psi_2(4)\,\mathfrak{F}\bar{\bar{P}}\Psi_1(1)\,\bar\Psi_1(2)\,\bar\Psi_2(3)\,\Psi_2(4)\;d\tau.$

(c) $\displaystyle\int \Psi_1(1)\,\Psi_2(2)\,\mathcal{G}\bar{\bar{P}}\Psi_1(1)\,\bar\Psi_2(2)\;d\tau.$

(d) $\displaystyle\int \Psi_1(1)\,\bar\Psi_2(2)\,\mathcal{G}\bar{\bar{P}}\bar\Psi_1(1)\,\Psi_2(2)\;d\tau.$

2. Calculate $G_{1212}{}^{\mathrm{MO}}$ (a) for allyl and (b) for ethylene. Here do not use actual value of repulsion integrals but instead use γ_{11}, γ_{12}, γ_{13}, etc. Also,

assume neglect of differential overlap. Remember that $\gamma_{11} = \gamma_{22}$, $\gamma_{12} = \gamma_{21}$, etc., in these cases.

3. What is the total electronic energy of benzene in its ground state if we assume Hückel MOs? Express this in terms of MO repulsion integrals and one-electron energies.

4. What is the total electronic energy of the allyl anion expressed in terms of AO integrals? Assume the Hückel eigenfunctions as an approximation.

5. Write the Slater determinant for the ground-state configuration of butadiene and then obtain its energy in terms of MO repulsion integrals.

6. Given a set of one-electron (e.g., Hückel) MOs for butadiene, write the wavefunction for its first excited singlet (i.e., S_1). Then write the wavefunctions for the lowest energy triplet (i.e., T_1); there should be three.

7. Consider three configurations of cyclobutadiene, ϕ_1, ϕ_2, and ϕ_3, where MOs 1 and 2 are filled, MOs 1 and 3 are filled, or where the usual excited singlet form is used with one electron in each of MOs 2 and 3 and two electrons in MO 1. Write the 3×3 matrix for interaction of the three configurations and then obtain final energies in terms of one-electron integrals. For simplicity and as an approximation treat the problem as if MO 1 and its electrons were absent. Note that the problem is easier if one uses the linear combinations $\psi_2 = (1/\sqrt{2})(\chi_1 - \chi_3)$ and $\psi_3 = (1/\sqrt{2})(\chi_2 - \chi_4)$ for MOs 2 and 3. Show that the results are independent of the choice for these MOs.

References

1. C. C. J. Roothaan, *Rev. Mod. Phys.* **23**, 69 (1951).
2. R. Daudel, R. Lefebvre, and C. Moser, "Quantum Chemistry, Methods and Applications." Wiley (Interscience), New York, 1961.
3. R. B. Parr, "The Quantum Theory of Molecular Electronic Structure." Benjamin, New York, 1963.
4. J. Pople and D. L. Beveridge, "Approximate Molecular Orbital Theory." McGraw-Hill, New York, 1970.
5. M. J. S. Dewar, "The Molecular Orbital Theory of Organic Chemistry." McGraw-Hill, New York, 1969.

ANSWERS TO PROBLEMS

Chapter 1

1.
$$\lim_{p \to \infty}[k^{5/2}/(3\pi)^{1/2}]\rho e^{-k\rho} = \lim_{p \to \infty}[k^{5/2}/(3\pi)^{1/2}](\rho/e^{k\rho})$$
$$= \lim_{p \to \infty}[k^{5/2}/(3\pi)^{1/2}](1/ke^{k\rho}) = 0.$$

2. Since for any point off the z axis $P(x, y, z)$, ρ is smaller for the corresponding point $P(0, 0, z)$, χ_z is maximized somewhere on the z axis. Here

$$\chi_z = (k^{5/2}/\sqrt{\pi})ze^{-kz}$$

since then $p = z$.

$$d\chi_z/dz = (k^{5/2}/\sqrt{\pi})(e^{-kz} - kze^{-kz}) = 0$$

which gives $z = 1/k$. [*Note:* We simplify the work by noting that when χ_z^2 is maximized, χ is a minimum or a maximum.]

3.
$$\begin{vmatrix} X & 1 & 0 & 0 & 0 & 1 \\ 1 & X & 1 & 0 & 0 & 0 \\ 0 & 1 & X & 1 & 0 & 0 \\ 0 & 0 & 1 & X & 1 & 0 \\ 0 & 0 & 0 & 1 & X & 1 \\ 1 & 0 & 0 & 0 & 1 & X \end{vmatrix}$$

4. $X = -\sqrt{3}, 0, 0, +\sqrt{3}$.

5. Using row 1,

$$A_{11} = \begin{vmatrix} X & 1 & 1 \\ 1 & X & 0 \\ 1 & 0 & X \end{vmatrix} = X^3 - 2X, \qquad A_{12} = - \begin{vmatrix} 1 & 1 & 1 \\ 0 & X & 0 \\ 0 & 0 & X \end{vmatrix} = -X^2$$

$$A_{13} = \begin{vmatrix} 1 & X & 1 \\ 0 & 1 & 0 \\ 0 & 1 & X \end{vmatrix} = X, \qquad A_{14} = - \begin{vmatrix} 1 & X & 1 \\ 0 & 1 & X \\ 0 & 1 & 0 \end{vmatrix} = X$$

for $X = 0$ all cofactors equal 0 and the ratios are indeterminate. The same situation results where cofactors of row 2 or of row 3 are used.

6.

$$\begin{array}{c} \\ \chi_1 \\ \chi_2 \\ \chi_3 \end{array} \begin{array}{ccc} \chi_1 & \chi_2 & \chi_3 \\ \begin{vmatrix} X & 0 & 0 \\ 0 & X & 1 \\ 0 & 1 & X \end{vmatrix} \end{array} = 0$$

Here elements A_{12} and A_{21} are orbitals; χ_1 and χ_2 are perpendicular and do not interact. Expansion by cofactors gives

$$X \begin{vmatrix} X & 1 \\ 1 & X \end{vmatrix} = 0 \quad \text{or} \quad X = 0 \quad \text{and} \quad \begin{vmatrix} X & 1 \\ 1 & X \end{vmatrix} = 0,$$

corresponding to an AO secular determinant $|X|$ and a separate ethylenic 2×2 secular determinant.

The ability of a secular determinant to be broken down depends on there being all zeros outside the blocks of the subdeterminants and no elements simultaneously in the rows or columns of more than one of the resulting subdeterminants (i.e., "block diagonalization").

7. $X = -1, -1, +2$. The array of energies is inverted from those of ordinary cyclopropenyl. For a closed shell, twist-hydrotrimethylenemethane should have four electrons in two orbitals and the anion is the preferred species. For cyclopropenyl, two electrons give a closed shell and the cation is favored.

8. $\begin{vmatrix} X & \cos\theta \\ \cos\theta & X \end{vmatrix} = 0, \qquad X^2 = \cos^2\theta, \qquad X = \pm\cos\theta.$

For $0°$, $X = -1, +1$; for $30°$, $X = -0.866, +0.866$; for $60°$, $X = -0.5$, $+0.5$; for $90°$, $X = 0, 0$; for $120°$, $X = +0.5, -0.5$; for $150°$, $X = +0.866$, -0.866; for $180°$, $X = +1, -1$.

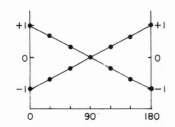

$X_a = -\cos\theta, \qquad \dfrac{dX_a}{d\theta} = \sin\theta.$

For $90°$, slope $= 1$.
Thus, twisting is forbidden.

9. The number of MOs equals the number of basis AOs. The number of basis AOs determines the order of the secular determinant and the latter determines the number of solutions.

Ethylene, methyl, and allyl have symmetrically disposed MO energy levels. This results in alternant hydrocarbons where alternating atoms may be designated as the "starred set" without two starred atoms or two unstarred atoms being adjacent.

10. $X = -1.618, -0.618, +0.618, +1.618.$

11.

$$\begin{array}{c c c c} & \chi_1 & \chi_2 & \chi_3' \\ \chi_1 & \begin{vmatrix} X & 1 & -1 \\ \\ \chi_2 & 1 & X & -1 \\ \\ \chi_3' & -1 & -1 & X \end{vmatrix} = 0 \end{array}$$

where χ_3' is χ_3 inverted. Multiplying column 3 and row 3 by -1, we obtain

$$\begin{array}{c c c c} & \chi_1 & \chi_2 & -\chi_3' \\ \chi_1 & \begin{vmatrix} X & 1 & 1 \\ \\ \chi_2 & 1 & X & 1 \\ \\ -\chi_3' & 1 & 1 & X \end{vmatrix} = 0 \end{array}$$

which affords the usual solutions for cyclopropenyl. However, we note that the heading now is $-\chi_3'$ which we see is indeed equal to χ_3.

Alternatively, the secular determinant in terms of χ_1, χ_2, and χ_3' could have been expanded to give the usual cyclopropenyl solutions.

12.

$$
D = \begin{array}{c|cccc}
 & \chi_1 & \chi_2 & \chi_3 & \chi_4 \\
\hline
\chi_1 & X & 1 & 0 & -1 \\
\chi_2 & 1 & X & 1 & 0 \\
\chi_3 & 0 & 1 & X & 1 \\
\chi_4 & -1 & 0 & 1 & X
\end{array} = 0
$$

$$
D = X \begin{vmatrix} X & 1 & 0 \\ 1 & X & 1 \\ 0 & 1 & X \end{vmatrix} - \begin{vmatrix} 1 & 1 & 0 \\ 0 & X & 1 \\ -1 & 1 & X \end{vmatrix} + \begin{vmatrix} 1 & X & 1 \\ 0 & 1 & X \\ -1 & 0 & 1 \end{vmatrix}
$$

$$
= X^4 - 4X^2 + 4, \quad (X^2 - 2)^2 = 0, \quad X = -\sqrt{2}, -\sqrt{2}, +\sqrt{2}, +\sqrt{2}
$$

The answers are the same as those from the circle mnemonic. Thus, Möbius cyclobutadiene is more stable.

13. For MO 1, $p_{14} = 0.137$ and $\Delta X = -2(0.137) = -0.276$ $X = -1.618 - 0.276 = -1.894$ (versus -2 as the exact solution). For MO 2, $p_{14} = -0.36$; $\Delta X = +0.72$; X then becomes $-0.618 + 0.72 = +0.10$ versus 0 for the exact solution. MO 3 gives $X = -0.10$ and MO 4 gives $+1.894$.

14. For MO 1, $p_{13} = 0.224$ and $\Delta X = -0.448$; thus $X = -1.62 - 0.45 = -2.07$. For MO 2, $p_{13} = -0.224$, $\Delta X = +0.448$, and $X = -0.618 + 0.448 = -0.17$. For MO 3, $p_{13} = -0.224$ and $X = 1.07$. For MO 4, $X = 1.17$. Compare the exact solutions $X = -2.17, -0.31, +1, +1.48$.

15. For MO 1, $p_{14} = -0.136$ and $\Delta X = +0.276$, giving $X = -1.62 + 0.28 = -1.34$. For MO 2, $p_{14} = +0.36$, $\Delta X = -0.72$, $X = -0.62 - 0.72 = -1.34$. For MO 3, $p_{14} = -0.36$ and $X = +1.34$. For MO 4, $X = 1.34$. The exact solutions are $-1.41, -1.41, +1.41, +1.41$.

16. The methoxyl stabilizes heavily the cationic system, moderately the radical, and does not stabilize the carbanion (note Chapter 4 for overlap destabilization effects indicating more than lack of stabilization for the anion).

We predict this since two adjacent (carbon) orbitals are nonbonding when separate but ethylenic (e.g., $X = \pm 1$) when overlapping. In the cation with an adjacent electron pair, the two electrons occupy the bonding MO giving stabilization. In the radical, there are three electrons, one is antibonding ($X = +1$) but two are bonding ($X = -1$) and net stabilization results. In the carbanion the bonding energy of two electrons is canceled by the antibonding energy of the remaining two electrons.

Chapter 3

1. For the benzene problem use one of the following approaches:

(a) Use a plane of symmetry σ_{14} and the symmetric group orbitals χ_1, $\chi_2 + \chi_6$, $\chi_3 + \chi_5$, χ_4 to give a 4×4 and the antisymmetric orbitals $\chi_2 - \chi_6$, $\chi_3 - \chi_5$ to give a 2×2.

(b) Use a plane of symmetry through bonds 2–3 and 5–6 and the symmetric group orbitals $\chi_1 + \chi_4$, $\chi_2 + \chi_3$, $\chi_5 + \chi_6$ to give a 3×3 as well as the antisymmetric orbitals $\chi_1 - \chi_4$, $\chi_2 - \chi_3$, $\chi_5 - \chi_6$ to give another 3×3.

(c) Use both planes of symmetry to get two 1×1's and two 2×2's.

(d) Use the C_{6v} or the D_{6h} group tables and formal group theory.

For the remaining molecules use the appropriate symmetry and parallel procedures.

2. Admix the three bonding MOs of ethylene in a Möbius fashion. This centers the circle at -1 (i.e., the energy of an isolated ethylenic orbital) and with a radius of 2ϵ where $-\epsilon$ is the resonance integral between transannular orbitals. Then use the same method with the antibonding MOs of ethylene with the circle centered at $+1$ and with a radius again of 2ϵ. All of the bonding ethylenic orbitals are orthogonal to all of the antibonding orbitals and thus the two problems can be treated separately.

Alternatively, use the circle device to do twist-hydrotrimethylene-methane. Then take each MO from one such twist-hydrotrimethylene-methane and admix it in a 2×2 with the corresponding MO of a second such moiety.

3. Use two planes of symmetry perpendicular to the molecule and informal group theory or instead the C_{2v} group. One obtains two 2×2's and two 3×3's. The eigenvalues are

$$-\tfrac{1}{2} \pm \tfrac{1}{2}(5)^{1/2}, \quad +\tfrac{1}{2} \pm \tfrac{1}{2}(5)^{1/2}, \quad -\tfrac{1}{2} \pm \tfrac{1}{2}(13)^{1/2}, \quad +\tfrac{1}{2} \pm \tfrac{1}{2}(13)^{1/2}, \quad -1, \quad +1$$

4 and 5. Three of the five MOs are norbornadiene ones which have the wrong symmetry to admix with the C-7 p orbital: These are

$$\tfrac{1}{2}(\chi_2 + \chi_3 + \chi_5 + \chi_6), \quad X = -1 - \epsilon$$

$$\tfrac{1}{2}(\chi_2 - \chi_3 - \chi_5 + \chi_6), \quad X = +1 - \epsilon$$

$$\tfrac{1}{2}(\chi_2 - \chi_3 + \chi_5 - \chi_6), \quad X = +1 + \epsilon$$

The remaining norbornadiene MOs, $\frac{1}{2}(\chi_2 + \chi_3 - \chi_5 - \chi_6)$ and χ_7, admix in a 2×2 to give

$$X = -\frac{1-\epsilon}{2} \pm \frac{1}{2}[(1-\epsilon)^2 + 16\epsilon^2]^{1/2}$$

Without χ_7, this last norbornadiene MO has an energy of $-1 + \epsilon$. The delocalization energy for the norbornadienyl cation is a function of ϵ, with stabilization increasing with overlap ϵ, while for the anion the DE is independent of ϵ.

6. $X = -1 - \epsilon,\ -1 - \epsilon,\ -1 + 2\epsilon,\ +1 - \epsilon,\ +1 - \epsilon,\ +1 + 2\epsilon.$

7. Pre- and postmultiplication by the vector $[1 \ -1]$ and its column transpose is equivalent to subtraction of the rows and then the columns. Pre- and postmultiplication by the two 2×2's is equivalent to addition and subtraction of rows and columns of the ethylene secular determinant and diagonalizes it.

8. Any second plane used must be perpendicular to the first. Otherwise, orbitals obtained with use of one plane will not be symmetry eigenfunctions relative to the other plane.

9. Addition of all rows gives one row all of whose elements are $X + 2$. This allows factorization of $(X + 2)$ from the secular determinant and gives the energy of the lowest MO (i.e., $X = -2$).

10.
$$\int \Psi_1 \Psi_2 \, d\tau = \int \left(\frac{1}{2}\chi_1 + \frac{1}{\sqrt{2}}\chi_2 + \frac{1}{2}\chi_3\right)\left(\frac{1}{\sqrt{2}}\chi_1 - \frac{1}{\sqrt{2}}\chi_3\right) d\tau$$

$$= \frac{1}{2\sqrt{2}}\int \chi_1^2 \, d\tau - \frac{1}{2\sqrt{2}}\int \chi_3^2 \, d\tau + \frac{1}{2}\int \chi_1\chi_2 \, d\tau - \frac{1}{2}\int \chi_2\chi_3 \, d\tau$$

$$= \frac{1}{2\sqrt{2}} - \frac{1}{2\sqrt{2}} + \frac{1}{2}S_{12} - \frac{1}{2}S_{23} = 0$$

11.
$$E = \int \left(\frac{1}{2}\chi_1 + \frac{1}{\sqrt{2}}\chi_2 + \frac{1}{2}\chi_3\right) \mathcal{H} \left(\frac{1}{2}\chi_1 + \frac{1}{\sqrt{2}}\chi_2 + \frac{1}{2}\chi_3\right) d\tau$$

$$= \frac{1}{4}\int \chi_1\mathcal{H}\chi_1 \, d\tau + \frac{1}{2}\int \chi_2\mathcal{H}\chi_2 \, d\tau + \frac{1}{4}\int \chi_3\mathcal{H}\chi_3 \, d\tau + \frac{1}{\sqrt{2}}\int \chi_1\mathcal{H}\chi_2$$

$$+ \frac{1}{\sqrt{2}}\int \chi_2\mathcal{H}\chi_3 \, d\tau = \alpha + \sqrt{2}\beta$$

or $X = -\sqrt{2}.$

12. This is equivalent to addition and subtraction of rows 1 and 3 and columns 1 and 3.

13. The group orbitals used are SS: $\chi_1 + \chi_6$, $\chi_2 + \chi_5 + \chi_7 + \chi_{10}$, $\chi_3 + \chi_4 + \chi_8 + \chi_9$; S_hA_v: $\chi_1 - \chi_6, \chi_2 - \chi_5 - \chi_7 + \chi_{10}, \chi_3 - \chi_4 - \chi_8 + \chi_9$; A_hS_v: $\chi_2 + \chi_5 - \chi_7 - \chi_{10}$, $\chi_3 + \chi_4 - \chi_8 - \chi_9$; AA: $\chi_2 - \chi_5 + \chi_7 - \chi_{10}$, $\chi_3 + \chi_4 - \chi_8 - \chi_9$. Solution of the two 3×3's and two 2×2's gives as eigenvalues $X = -2.308$, -1.618, -1.303, -1.000, -0.618, $+0.618$, $+1.000$, $+1.303$, $+1.618$, $+2.308$.

We have encountered the combinations involving $\chi_2, \chi_3, \chi_4, \chi_5, \chi_7, \chi_8, \chi_9$, and χ_{10} alone in the butadiene problem. The MOs are just sums and differences of two butadiene moiety MOs.

14. (a) Use two perpendicular planes of symmetry and transannular overlap of ϵ. One obtains $X = \pm 1, -\frac{1}{2} \pm \frac{1}{2}(5)^{1/2}, \frac{1}{2} \pm \frac{1}{2}(5 + 16\epsilon^2)^{1/2}$.

(b) Use two perpendicular planes of symmetry or the D_{2d} group table to obtain $X = -1, -1, 1 \pm 2\epsilon$.

(c) Degenerate pairs at $X = -\frac{1}{2} \pm \frac{1}{2}(5)^{1/2}$, also $X = \frac{1}{2} - \epsilon \pm \frac{1}{2}(4\epsilon^2 + 4\epsilon + 5)^{1/2}$, $X = +\frac{1}{2} + \epsilon \pm \frac{1}{2}(4\epsilon^2 - 4\epsilon + 5)^{1/2}$. [*Note:* Picking the basis set orientation so that there is only plus–plus or minus–minus overlap, convenient symmetry orbitals are $\chi_1 + \chi_4$ mixing with $\chi_2 + \chi_3$, $\chi_5 + \chi_8$ mixing with $\chi_6 + \chi_7$, $\chi_1 - \chi_4 + \chi_5 - \chi_8$ mixing with $\chi_2 - \chi_3 + \chi_6 - \chi_7$, and $\chi_1 - \chi_4 - \chi_5 + \chi_8$ mixing with $\chi_2 - \chi_3 - \chi_6 + \chi_7$.]

(d) For molecule (a) the bonding ethylenic MO does not have the right symmetry to mix with any butadiene MOs. The antibonding ethylenic MO has the right symmetry to admix with Ψ_2 and Ψ_4 of butadiene. Ψ_1 and Ψ_3 of butadiene do not admix with any other MO. The reasoning for molecules (b) and (c) is similar; in each case MOs symmetric in one ring will not mix with any MOs of the other ring.

15. Using two planes of symmetry and the numbering

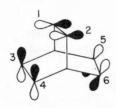

the group orbitals which are totally antisymmetric are $\chi_1 - \chi_2$ and $\chi_3 - \chi_4 + \chi_5 - \chi_6$. This gives $1 + 2\epsilon$ and $1 - \epsilon$ as solutions. The SS orbital is $\chi_3 + \chi_4 - \chi_5 - \chi_6$ $(X = -1 - \epsilon)$. The SA orbital is $\chi_3 - \chi_4 - \chi_5 + \chi_6$ $(X = 1 - \epsilon)$. The AS orbitals are $\chi_1 + \chi_2$ and $\chi_3 + \chi_4 + \chi_5 + \chi_6$ $(X = -1 + 2\epsilon$ and $-1 - \epsilon)$.

16. The p orbital (χ_7) has the correct symmetry to admix only with $\chi_2 + \chi_3 - \chi_5 - \chi_6$, where χ_2, χ_3, χ_5, and χ_6 are chosen all in the same direction. The final MO energies are $1 + \epsilon$, $1 - \epsilon$, $-1 - \epsilon$, $-\frac{1}{2} + (\epsilon/2) \pm \frac{1}{2}(1 - 2\epsilon + 17\epsilon^2)^{1/2}$.

17. The eigenfunctions consist of normalized $\pm$ combinations of the separate benzene MOs. The eigenvalues are $X = -2 \pm \epsilon$, $-1 \pm \epsilon$, $-1 \pm \epsilon$, $+1 \pm \epsilon$, $+1 \pm \epsilon$, and $+2 \pm \epsilon$. This problem is really more readily done by admixing corresponding benzene MOs from the two rings in a series of 2×2's. There is no additional delocalization energy (i.e., DE is independent of ϵ) until ϵ reaches 1.

18. (a) The multiplication gives $2h$ since $\tilde{\mathbf{V}}_{A1}\mathbf{V}_{A1} = h$ but all other products are zero due to orthogonality. After division by h, we get 2 which is the number of times the irreducible vector $\mathbf{V}_{A1}$ occurs in $\mathbf{V}_r$. This gives us a way to determine how an irreducible vector (or representation) is contained in a reducible one.

(b) Note Eqs. 3.5-4 through 3.5-6.

19. (a) One 2×2 mixes $\chi_1 + \chi_4$ and $\chi_2 + \chi_3$; the other mixes $\chi_1 - \chi_4$, $\chi_2 - \chi_3$. There is a crossing of MOs at $\theta = 45°$. Using geometric symmetry alone, one might have predicted no crossing since the MOs crossing have the same ordinary symmetry. But the symmetry element used (i.e., a C_2 axis) does not go through any bonds which are changing and is not useful. Actually, there is a "hidden symmetry" since the four-orbital array is cyclobutadienoid and the crossing orbitals actually do differ if considered using cyclobutadienoid symmetry; crossing does occur as the explicit calculation shows.

(b) One obtains the same result. Any other linear combination of the two central basis orbitals is also acceptable.

(c) In 19a with $\theta = 45°$ we have Hückel system, χ_1 overlaps with χ_2, which overlaps with χ_3, which overlaps with χ_4, which overlaps with χ_1—all plus–plus.

In 19b with initial geometry the overlap is χ_1 with χ_2 with χ_3 with χ_4 with χ_1 but there is an odd number of plus–minus overlaps (e.g., χ_1 with χ_4).

20. The six $(1/\sqrt{2})(\chi_a + \chi_b)$ sets can be mixed as can the six $(1/\sqrt{2})(\chi_a - \chi_b)$ sets, each in benzenoid fashion. The MOs are exactly analogous to benzene MOs except here we have linear combinations of $(1/\sqrt{2})(\chi_a + \chi_b)$ and $(1/\sqrt{2})(\chi_a - \chi_b)$.

21. Here we mix each twist-hydrotrimethylenemethane MO with the same MO from a second molecule oriented face to face. Nonequivalent MOs do not mix due to different symmetries. The eigenvalues and eigenfunctions are the same as obtained previously in more traditional but more tedious fashion.

22. This is done somewhat analogously to Problem 17 except there we used three planes of symmetry rather than formal group theory. Here the C_{2v} group can be used.

23. Following the barrelene treatment in Problem 21, we mix corresponding MOs of two cyclobutadiene molecules in a series of 2×2's. The energies are $-2 \pm \epsilon$, $0 \pm \epsilon$, $0 \pm \epsilon$, $+2 \pm \epsilon$. Again the justification for mixing only corresponding MOs is the failure of MOs of different symmetry to mix; this can be tested via a secular determinant.

24. Once. Use Rule I. Thus transpose $[1 \ -1 \ -1 \ 1][4 \ 0 \ -2 \ -2]^{\frac{1}{4}} = 1$.

25. $X = 0, \dfrac{1}{\sqrt{2}}(\chi_2 - \chi_4); \quad X = 1, \dfrac{1}{\sqrt{2}}(\chi_1 - \chi_3); \quad X = -\dfrac{1}{2} \pm \dfrac{1}{2}(17)^{1/2}.$

Chapter 4

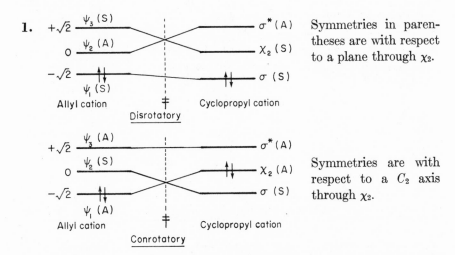

1.

Allyl cation — Disrotatory — Cyclopropyl cation

Symmetries in parentheses are with respect to a plane through χ_2.

Allyl cation — Conrotatory — Cyclopropyl cation

Symmetries are with respect to a C_2 axis through χ_2.

2. The results are the same. The correlation lines are obtained by noting the Hückel cyclopropenyl array of MOs for the disrotatory closure at the transition state; the degeneracy at $+1$ shows MOs ψ_2 and ψ_3 cross. For conrotatory closure a Möbius cyclopropenyl array of MOs is used to determine transition state MO ordering and the -1 degeneracy shows MOs ψ_1

and ψ_2 cross. The transition state point along the reaction coordinate is shown above in the answer to Problem 1 by a ‡.

3. Refer to Fig. 1.7-A; however, note that the 1,3-bond order is $\epsilon_{13}c_{1k}c_{3k}$ for each MO k and that ϵ will be larger than the ± 1 used for ordinary π overlap. Also for top–top (Hückel) closure ϵ is positive and for top–bottom (Möbius) overlap ϵ is negative. Take $\epsilon = +2$ or -2 for two cases (note overlap for a σ bond will be stronger than for a π bond; hence $\epsilon = \pm 2$ rather than ± 1 is reasonable). Then we get the correlation diagrams:

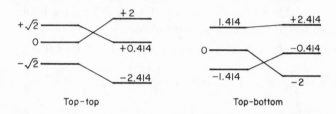

Top-top   Top-bottom

4. This treatment is similar to that of Problem 13 in Chapter 1 except that, again, we use a large value of ϵ to get the perturbation energy (i.e., ΔX) for each MO on initiation of 1,4-sigma bonding. For conrotatory twisting MOs 1 and 2 cross as do 3 and 4. For disrotatory twisting MOs 2 and 3 cross. Use $\epsilon = +2$ for Hückel closure and $\epsilon = -2$ for Möbius closure.

5 and 6. At the half-way stage, we have "square cyclobutadiene" and a degeneracy at zero. The correlation diagram is

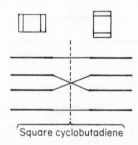

Square cyclobutadiene

The interconversion of two "cyclohexatriene tautomers" goes through a Hückel, hexagonal system with no degeneracies at zero and is allowed. All the $4N$ systems have nonbonding degeneracies and are forbidden while the $4N + 2$ do not have such degeneracies and are allowed. We predict Jahn–Teller molecular distortion of the $4N$ systems.

7. (a) Hückel, $4e$, forbidden; (b) Möbius, $4e$, allowed; (c) Möbius, $4e$, allowed; (d) Hückel, $4e$, forbidden; (e) Hückel, $4e$, forbidden.

8.

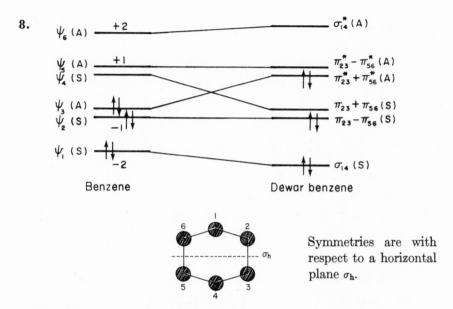

Benzene   Dewar benzene

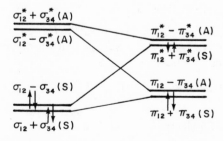

Symmetries are with respect to a horizontal plane σ_h.

The ground-state reaction is forbidden.

The MOs which are antisymmetric with respect to a plane bisecting C-1 and C-4 are ψ_2 (becoming $\pi_{23} - \pi_{56}$) and ψ_5 (becoming $\pi_{23}^* - \pi_{56}^*$). These do not cross zero and ψ_2 remains doubly occupied and bonding throughout; these do not affect the forbiddenness. The remaining four MOs are made up of a Hückel array of *group* orbitals and have $4e$'s. Thus, we have a forbidden reaction. For such bicyclic transition states, consider a single ring and localize the remaining electrons in the other with no double bonds at the overlapping carbons. The single ring then becomes determining.

9. (a) 4/7 benzylic; (b) 9/20 α-naphthylic. Benzylic more basic.

10. Symmetries are relative to a plane bisecting the bonds cleaving.

The reaction is forbidden.

Chapter 5

1. (a) Zero, spin orthogonality for electrons 3 as well as 4.
    (b) Nonzero since on permutation we obtain

$$- \int \Psi_1(1)\bar{\Psi}_1(2)\Psi_2(3)\bar{\Psi}_2(4)\,\mathcal{F}\Psi_1(1)\bar{\Psi}_1(2)\Psi_2(3)\bar{\Psi}_2(4)\ d\tau$$

$$= -2I_1 - 2I_2 - G_{1111}{}^{\text{MO}} - 4G_{1212}{}^{\text{MO}} - G_{2222}{}^{\text{MO}}$$

(c) Zero, spin orthogonality.
(d) Nonzero. Permutation gives $G_{1221}{}^{\text{MO}}$.

2. (a) $\frac{1}{4}\gamma_{11} + \frac{1}{2}\gamma_{12} + \frac{1}{4}\gamma_{13}$.   (b) $\frac{1}{2}\gamma_{11} + \frac{1}{2}\gamma_{12}$.

3. $E = 2I_1 + 2I_2 + 2I_3 + G_{1111}{}^{\text{MO}} + G_{2222}{}^{\text{MO}} + G_{3333}{}^{\text{MO}} + 4G_{1212}{}^{\text{MO}}$

$$+ 4G_{2323}{}^{\text{MO}} + 4G_{1313}{}^{\text{MO}} - 2G_{1221}{}^{\text{MO}} - 2G_{1331}{}^{\text{MO}} - 2G_{2332}{}^{\text{MO}}.$$

4. $E = 2I_1 + 2I_2 + G_{1111}{}^{\text{MO}} + G_{2222}{}^{\text{MO}} + 4G_{1212}{}^{\text{MO}} - 2G_{1221}{}^{\text{MO}}$ where

$I_1 = -\sqrt{2}|\beta|$, $I_2 = 0|\beta|$, $G_{1111}{}^{\text{MO}} = \frac{3}{8}\gamma_{11} + \frac{1}{2}\gamma_{12} + \frac{1}{8}\gamma_{13}$,

$G_{2222}{}^{\text{MO}} = \frac{1}{2}\gamma_{11} + \frac{1}{2}\gamma_{13}$, $G_{1212}{}^{\text{MO}} = \frac{1}{4}\gamma_{11} + \frac{1}{2}\gamma_{12} + \frac{1}{4}\gamma_{13}$,   and

$G_{1221}{}^{\text{MO}} = \frac{1}{4}\gamma_{11} - \frac{1}{4}\gamma_{13}$. Thus $E = -2\sqrt{2}|\beta| + \frac{11}{8}\gamma_{11} + \frac{5}{2}\gamma_{12} + \frac{17}{8}\gamma_{13}$.

5. $\Phi = |\Psi_1(1)\bar{\Psi}_1(2)\Psi_2(3)\bar{\Psi}_2(4)|$ $\overset{N}{}$ (N implies normal-
    $E = 2I_1 + 2I_2 + G_{1111}{}^{\text{MO}} + 4G_{1212}{}^{\text{MO}} + G_{2222}{}^{\text{MO}}$ ization by $1/(4!)^{1/2}$)

6. $^1\Phi = \frac{1}{\sqrt{2}} \{ |\Psi_1(1)\bar{\Psi}_1(2)\Psi_2(3)\bar{\Psi}_3(4)|^N - |\Psi_1(1)\bar{\Psi}_1(2)\bar{\Psi}_2(3)\Psi_3(4)|^N \}$

$$^3\Phi_a = \frac{1}{\sqrt{2}} \{ |\Psi_1(1)\bar{\Psi}_1(2)\Psi_2(3)\bar{\Psi}_3(4)|^N + |\Psi_1(1)\bar{\Psi}_1(2)\bar{\Psi}_2(3)\Psi_3(4)|^N \}$$

$$^3\Phi_b = |\Psi_1(1)\bar{\Psi}_1(2)\Psi_2(3)\Psi_3(4)|^N \quad \text{and} \quad ^3\Phi_c = |\Psi_1(1)\bar{\Psi}_1(2)\bar{\Psi}_2(3)\bar{\Psi}_3(4)|^N .$$

7. $F_{11} = 2I_2 + G_{2222}{}^{\text{MO}}$, $F_{22} = 2I_3 + G_{3333}{}^{\text{MO}}$, $F_{33} = I_2 + I_3 + G_{2323} + G_{2332}$,
$F_{12} = G_{2233}$, $F_{13} = \sqrt{2}G_{2223}$, $F_{23} = \sqrt{2}G_{3332}$. All off-diagonal elements vanish with this choice of a basis.

$$G_{2222}{}^{\text{MO}} = G_{3333}{}^{\text{MO}} = \frac{1}{2}\gamma_{11} + \frac{1}{2}\gamma_{13}. \qquad G_{2323}{}^{\text{MO}} = \gamma_{12}. \ G_{2332}{}^{\text{MO}} = 0$$

Thus $E_3 = 2I + \gamma_{12}$, $E_2 = 2I + \frac{1}{2}\gamma_{11} + \frac{1}{2}\gamma_{13} = E_1$,   where I is the one-electron energy of an electron in MO 2 or 3.

INDEX

B 5
C 6
D 7
E 8
F 9
G 0
H 1
I 2
J 3